Django

3.0 入门与实践

李健 编著

清华大学出版社
北京

内 容 简 介

本书是一线程序员多年开发经验的结晶。书中深入浅出地讲解了使用 Django 开发 Web 网站所需的配置、后台、路由系统、模型、视图、模板、表单系统、中间件、自动化测试、国际化及本地化、安全和部署等内容，帮助读者快速进入 Web 项目开发，在项目实践中灵活应用各种开发技术和方法。

本书主要包含四部分：第一部分（第 1 章）为读者介绍 Django 的发展状况以及如何搭建 Django 开发环境；第二部分（第 2 章）主要介绍什么是 Web 开发框架并通过搭建一个简单的 Web 框架帮助读者理解 Web 框架的工作原理；第三部分（第 3 章）带领读者搭建一个网站，使读者能够从整体上了解 Django；第四部分（第 4～16 章）详细介绍 Django 各个模块的工作原理，通过学习这部分内容，读者将能够独立开发 Django 应用。

本书可供 Web 开发初中级读者以及希望使用 Python 作为编程语言的软件开发工程师参考。

本书封面贴有清华大学出版社防伪标签，无标签者不得销售。
版权所有，侵权必究。举报：010-62782989，beiqinquan@tup.tsinghua.edu.cn。

图书在版编目 (CIP) 数据

Django 3.0 入门与实践 / 李健编著 . —北京：清华大学出版社，2021.1
ISBN 978-7-302-56714-1

Ⅰ. ①D… Ⅱ. ①李… Ⅲ. ①软件工具—程序设计 Ⅳ. ① TP311.561

中国版本图书馆 CIP 数据核字 (2020) 第 203512 号

责任编辑：秦　健
封面设计：杨玉兰
责任校对：胡伟民
责任印制：丛怀宇

出版发行：清华大学出版社
网　　址：http://www.tup.com.cn，http://www.wqbook.com
地　　址：北京清华大学学研大厦 A 座　　　　邮　　编：100084
社 总 机：010-62770175　　　　　　　　　　邮　　购：010-83470235
投稿与读者服务：010-62776969，c-service@tup.tsinghua.edu.cn
质 量 反 馈：010-62772015，zhiliang@tup.tsinghua.edu.cn

印 装 者：三河市中晟雅豪印务有限公司
经　　销：全国新华书店
开　　本：186mm×240mm　　　印　张：19.75　　　字　数：560 千字
版　　次：2021 年 2 月第 1 版　　印　次：2021 年 2 月第 1 次印刷
定　　价：79.00 元

产品编号：088856-01

Foreword 前　言

非常感谢你选择本书，希望通过阅读本书能够为你带来帮助。

相信你在选择本书的时候已经对 Django 有了一定的了解，但是还请允许我在这里继续为大家介绍一下 Django 以及为什么写作本书。Django 是用 Python 语言开发的一套开源 Web 框架，Python 语言作为目前最流行的编程语言之一，已经不仅仅满足于早期的脚本开发，它被越来越多地应用于大型的、前沿的项目中，如科学计算和人工智能等。Django 的出现为众多 Python 爱好者带来了福音，也为程序员的职业生涯拓宽了道路，现在我们也可以使用 Python 语言开发 Web 应用了。笔者有过多年的 Web 开发经验以及持续集成系统开发经验，曾经使用 ASP.NET 开发过 Web 应用，也用 Jenkins、Shell 开发过持续集成系统，但是，当使用 Django 开发网站的时候才真正体会到将脚本语言应用到 Web 开发中的乐趣，尤其是在搭建个人网站时，那种快速、自由的编码体验真的令人兴奋。为了对自己的工作做一个总结，也为更多初学者提供一本参考手册，所以决定写作本书。

本书主要包含四部分：

第一部分（第 1 章）为读者介绍 Django 的发展状况以及如何搭建 Django 开发环境。

第二部分（第 2 章）主要介绍什么是 Web 开发框架并通过搭建一个简单的 Web 框架帮助读者理解 Web 框架的工作原理。

第三部分（第 3 章）带领读者搭建一个网站，使读者能够从整体上了解 Django。

第四部分（第 4 ～ 16 章）详细介绍 Django 各个模块的工作原理，通过学习这部分内容，读者将能够独立开发 Django 应用。

本书读者需要比较熟练地掌握 Python 语言，同时具备一定的 Web 开发基础，能够比较熟练地使用 HTML、CSS、JavaScript，最好掌握一定的数据库开发知识，以便能够更容易地理解 Django 的 ORM 模型。

本书非常适合准备转向 Web 开发的 Python 工程师阅读，也适合正在使用 Django 开发 Web 应用的工程师作为参考手册。

在编写本书的过程中参考了 Django 官方文档，在此对 Django 团队以及社区表示真心的感谢。另外，编写本书以及学习 Django 的过程中从 StackOverflow 社区也获得了大量帮助，在此表示由衷的感谢。

由于本人能力有限，在编写本书的过程中可能有所疏漏，敬请读者指正。如果读者在阅读过程中发现本书的问题或者有好的建议，欢迎通过清华大学出版社网站（www.tup.com.cn）或者扫描如右二维码联系我们。

李健

Contents 目 录

第 1 章 走进 Django 的世界 ⋯⋯⋯⋯⋯⋯ 1
- 1.1 认识 Django ⋯⋯⋯⋯⋯⋯⋯⋯⋯⋯⋯ 1
- 1.2 版本选择 ⋯⋯⋯⋯⋯⋯⋯⋯⋯⋯⋯⋯ 1
- 1.3 搭建开发环境 ⋯⋯⋯⋯⋯⋯⋯⋯⋯⋯ 2
  - 1.3.1 安装 Python ⋯⋯⋯⋯⋯⋯⋯⋯ 2
  - 1.3.2 安装 Django ⋯⋯⋯⋯⋯⋯⋯⋯ 2

第 2 章 Web 开发框架 ⋯⋯⋯⋯⋯⋯⋯⋯ 4
- 2.1 Socket 编程 ⋯⋯⋯⋯⋯⋯⋯⋯⋯⋯⋯ 4
- 2.2 MTV 框架 ⋯⋯⋯⋯⋯⋯⋯⋯⋯⋯⋯ 6

第 3 章 搭建第一个 Django 网站 ⋯⋯⋯ 8
- 3.1 创建项目 ⋯⋯⋯⋯⋯⋯⋯⋯⋯⋯⋯⋯ 8
- 3.2 运行项目 ⋯⋯⋯⋯⋯⋯⋯⋯⋯⋯⋯⋯ 9
- 3.3 创建投票应用 ⋯⋯⋯⋯⋯⋯⋯⋯⋯ 10
- 3.4 开发第一个视图 ⋯⋯⋯⋯⋯⋯⋯⋯ 11
- 3.5 配置数据库 ⋯⋯⋯⋯⋯⋯⋯⋯⋯⋯ 12
- 3.6 创建模型 ⋯⋯⋯⋯⋯⋯⋯⋯⋯⋯⋯ 13
- 3.7 激活模型 ⋯⋯⋯⋯⋯⋯⋯⋯⋯⋯⋯ 14
- 3.8 Django 管理页面 ⋯⋯⋯⋯⋯⋯⋯⋯ 15
- 3.9 向管理页面中添加投票应用 ⋯⋯ 16
- 3.10 添加视图 ⋯⋯⋯⋯⋯⋯⋯⋯⋯⋯ 18
  - 3.10.1 扩展视图 ⋯⋯⋯⋯⋯⋯⋯⋯ 19
  - 3.10.2 处理 404 错误 ⋯⋯⋯⋯⋯⋯ 21
- 3.11 使用模板系统 ⋯⋯⋯⋯⋯⋯⋯⋯ 22
  - 3.11.1 模板中的超链接 ⋯⋯⋯⋯⋯ 23
  - 3.11.2 为超链接添加命名空间 ⋯⋯ 23
- 3.12 HTML 表单 ⋯⋯⋯⋯⋯⋯⋯⋯⋯ 24
- 3.13 添加样式 ⋯⋯⋯⋯⋯⋯⋯⋯⋯⋯ 27
- 3.14 本地化 ⋯⋯⋯⋯⋯⋯⋯⋯⋯⋯⋯ 28
- 3.15 小结 ⋯⋯⋯⋯⋯⋯⋯⋯⋯⋯⋯⋯ 29

第 4 章 django-admin 和 manage.py ⋯⋯ 30
- 4.1 help ⋯⋯⋯⋯⋯⋯⋯⋯⋯⋯⋯⋯⋯ 31
- 4.2 version ⋯⋯⋯⋯⋯⋯⋯⋯⋯⋯⋯⋯ 32
- 4.3 check ⋯⋯⋯⋯⋯⋯⋯⋯⋯⋯⋯⋯ 32
- 4.4 compilemessages ⋯⋯⋯⋯⋯⋯⋯ 32
- 4.5 createcachetable ⋯⋯⋯⋯⋯⋯⋯⋯ 34
- 4.6 dbshell ⋯⋯⋯⋯⋯⋯⋯⋯⋯⋯⋯⋯ 34
- 4.7 diffsettings ⋯⋯⋯⋯⋯⋯⋯⋯⋯⋯ 35
- 4.8 dumpdata ⋯⋯⋯⋯⋯⋯⋯⋯⋯⋯⋯ 35
- 4.9 flush ⋯⋯⋯⋯⋯⋯⋯⋯⋯⋯⋯⋯⋯ 36
- 4.10 inspectdb ⋯⋯⋯⋯⋯⋯⋯⋯⋯⋯ 36
- 4.11 loaddata ⋯⋯⋯⋯⋯⋯⋯⋯⋯⋯⋯ 38
- 4.12 makemessages ⋯⋯⋯⋯⋯⋯⋯⋯ 38
- 4.13 startproject ⋯⋯⋯⋯⋯⋯⋯⋯⋯⋯ 39
- 4.14 startapp ⋯⋯⋯⋯⋯⋯⋯⋯⋯⋯⋯ 39
- 4.15 runserver ⋯⋯⋯⋯⋯⋯⋯⋯⋯⋯⋯ 40
- 4.16 sendtestemail ⋯⋯⋯⋯⋯⋯⋯⋯⋯ 40
- 4.17 shell ⋯⋯⋯⋯⋯⋯⋯⋯⋯⋯⋯⋯ 41
- 4.18 迁移 ⋯⋯⋯⋯⋯⋯⋯⋯⋯⋯⋯⋯ 41
  - 4.18.1 makemigrations ⋯⋯⋯⋯⋯⋯ 41
  - 4.18.2 migrate ⋯⋯⋯⋯⋯⋯⋯⋯⋯ 42
  - 4.18.3 sqlmigrate ⋯⋯⋯⋯⋯⋯⋯⋯ 43
  - 4.18.4 showmigrations ⋯⋯⋯⋯⋯⋯ 43
- 4.19 changepassword ⋯⋯⋯⋯⋯⋯⋯⋯ 44
- 4.20 createsuperuser ⋯⋯⋯⋯⋯⋯⋯⋯ 44
- 4.21 collectstatic ⋯⋯⋯⋯⋯⋯⋯⋯⋯ 44
- 4.22 findstatic ⋯⋯⋯⋯⋯⋯⋯⋯⋯⋯ 45
- 4.23 默认选项 ⋯⋯⋯⋯⋯⋯⋯⋯⋯⋯ 45

第 5 章 配置 ⋯⋯⋯⋯⋯⋯⋯⋯⋯⋯⋯ 46
- 5.1 Django 配置文件 ⋯⋯⋯⋯⋯⋯⋯⋯ 46

- 5.1.1 引用 Django 配置信息 ... 47
- 5.1.2 django.setup ... 47
- 5.2 Cache ... 48
  - 5.2.1 CACHES ... 48
  - 5.2.2 CACHE_MIDDLEWARE_ALIAS: ... 49
  - 5.2.3 CACHE_MIDDLEWARE_KEY_PREFIX: ... 49
  - 5.2.4 CACHE_MIDDLEWARE_SECONDS: ... 49
- 5.3 数据库 ... 50
  - 5.3.1 DATABASES ... 50
  - 5.3.2 DATABASE_ROUTERS ... 54
  - 5.3.3 DEFAULT_INDEX_TABLESPACE ... 57
  - 5.3.4 DEFAULT_TABLESPACE ... 57
- 5.4 调试 ... 57
  - 5.4.1 DEBUG ... 57
  - 5.4.2 DEBUG_PROPAGATE_EXCEPTIONS ... 58
- 5.5 电子邮件 ... 58
  - 5.5.1 ADMINS ... 58
  - 5.5.2 DEFAULT_FROM_EMAIL ... 59
  - 5.5.3 EMAIL_BACKEND ... 59
  - 5.5.4 EMAIL_HOST ... 59
  - 5.5.5 EMAIL_HOST_USER ... 59
  - 5.5.6 EMAIL_HOST_PASSWORD ... 59
  - 5.5.7 EMAIL_PORT ... 59
  - 5.5.8 EMAIL_TIMEOUT ... 59
  - 5.5.9 SERVER_EMAIL ... 59
  - 5.5.10 MANAGERS ... 60
- 5.6 文件上传 ... 60
  - 5.6.1 DEFAULT_FILE_STORAGE ... 60
  - 5.6.2 FILE_CHARSET ... 60
  - 5.6.3 FILE_UPLOAD_HANDLERS ... 60
  - 5.6.4 FILE_UPLOAD_MAX_MEMORY_SIZE ... 60
  - 5.6.5 FILE_UPLOAD_PERMISSIONS ... 60
  - 5.6.6 FILE_UPLOAD_DIRECTORY_PERMISSIONS ... 60
  - 5.6.7 FILE_UPLOAD_TEMP_DIR ... 61
  - 5.6.8 MEDIA_ROOT ... 61
  - 5.6.9 MEDIA_URL ... 61
  - 5.6.10 静态文件 ... 62
- 5.7 表单 ... 63
- 5.8 国际化（i18n/l10n）... 63
  - 5.8.1 DECIMAL_SEPARATOR ... 63
  - 5.8.2 NUMBER_GROUPING ... 63
  - 5.8.3 THOUSAND_SEPARATOR ... 65
  - 5.8.4 USE_THOUSAND_SEPARATOR ... 66
  - 5.8.5 FIRST_DAY_OF_WEEK ... 66
  - 5.8.6 DATE_FORMAT ... 66
  - 5.8.7 DATE_INPUT_FORMATS ... 66
  - 5.8.8 DATETIME_FORMAT ... 67
  - 5.8.9 SHORT_DATE_FORMAT ... 67
  - 5.8.10 SHORT_DATETIME_FORMAT ... 67
  - 5.8.11 DATETIME_INPUT_FORMATS ... 67
  - 5.8.12 TIME_FORMAT ... 68
  - 5.8.13 TIME_INPUT_FORMATS ... 68
  - 5.8.14 YEAR_MONTH_FORMAT ... 68
  - 5.8.15 MONTH_DAY_FORMAT ... 68
  - 5.8.16 TIME_ZONE ... 69
  - 5.8.17 LANGUAGE_CODE ... 69
  - 5.8.18 LANGUAGE_COOKIE_AGE ... 69
  - 5.8.19 LANGUAGE_COOKIE_DOMAIN ... 69
  - 5.8.20 LANGUAGE_COOKIE_NAME ... 70
  - 5.8.21 LANGUAGE_COOKIE_PATH ... 70
  - 5.8.22 LANGUAGES ... 70
  - 5.8.23 LANGUAGES_BIDI ... 70
  - 5.8.24 LOCALE_PATHS ... 70
  - 5.8.25 USE_I18N ... 70
  - 5.8.26 USE_L10N ... 71
  - 5.8.27 USE_TZ ... 71
  - 5.8.28 Python datetime 语法 ... 71
- 5.9 HTTP ... 71
  - 5.9.1 DATA_UPLOAD_MAX_MEMORY_SIZE ... 71
  - 5.9.2 DATA_UPLOAD_MAX_NUMBER_FIELDS ... 72
  - 5.9.3 DEFAULT_CHARSET ... 72
  - 5.9.4 DISALLOWED_USER_AGENTS ... 72

5.9.5	FORCE_SCRIPT_NAME	72
5.9.6	INTERNAL_IPS	72
5.9.7	SECURE_BROWSER_XSS_FILTER	73
5.9.8	SECURE_CONTENT_TYPE_NOSNIFF	73
5.9.9	SECURE_HSTS_INCLUDE_SUBDOMAINS	73
5.9.10	SECURE_HSTS_PRELOAD	73
5.9.11	SECURE_HSTS_SECONDS	73
5.9.12	SECURE_PROXY_SSL_HEADER	74
5.9.13	SECURE_REDIRECT_EXEMPT	74
5.9.14	SECURE_REFERRER_POLICY	75
5.9.15	SECURE_SSL_HOST	75
5.9.16	SECURE_SSL_REDIRECT	75
5.9.17	SIGNING_BACKEND	75
5.9.18	WSGI_APPLICATION	75
5.10	安全	75
5.10.1	SECRET_KEY	75
5.10.2	ALLOWED_HOSTS	76
5.11	CSRF	76
5.11.1	CSRF_COOKIE_AGE	76
5.11.2	CSRF_COOKIE_DOMAIN	77
5.11.3	CSRF_COOKIE_HTTPONLY	77
5.11.4	CSRF_COOKIE_NAME	77
5.11.5	CSRF_COOKIE_PATH	77
5.11.6	CSRF_COOKIE_SAMESITE	77
5.11.7	CSRF_COOKIE_SECURE	77
5.11.8	CSRF_USE_SESSIONS	78
5.11.9	CSRF_FAILURE_VIEW	78
5.11.10	CSRF_HEADER_NAME	78
5.11.11	CSRF_TRUSTED_ORIGINS	78
5.11.12	代码示例	78
5.12	模型	81
5.12.1	ABSOLUTE_URL_OVERRIDES	81
5.12.2	FIXTURE_DIRS	81
5.12.3	INSTALLED_APPS	81
5.13	日志	81
5.13.1	LOGGING	81
5.13.2	LOGGING_CONFIG	82
5.14	模板	82
5.15	URLs	83
5.15.1	ROOT_URLCONF	83
5.15.2	APPEND_SLASH	83
5.15.3	PREPEND_WWW	84
5.16	其他	84
5.16.1	DEFAULT_EXCEPTION_REPORTER_FILTER	84
5.16.2	MIDDLEWARE	84

第 6 章　后台管理页面 85

6.1	ModelAdmin 属性	85
6.1.1	date_hierarchy	86
6.1.2	actions_on_top/actions_on_bottom	88
6.1.3	actions_selection_counter	88
6.1.4	empty_value_display	89
6.1.5	exclude	90
6.1.6	fields	91
6.1.7	fieldsets	92
6.1.8	filter_horizontal	94
6.1.9	filter_vertical	95
6.1.10	form	96
6.1.11	formfield_overrides	96
6.1.12	inlines	97
6.1.13	list_display	97
6.1.14	list_display_links	100
6.1.15	list_editable	101
6.1.16	list_filter	102
6.1.17	list_per_page	102
6.1.18	list_max_show_all	103
6.1.19	list_select_related	103
6.1.20	ordering	104
6.1.21	paginator	104
6.1.22	prepopulated_fields	104
6.1.23	preserve_filters	104
6.1.24	radio_fields	105
6.1.25	autocomplete_fields	105
6.1.26	raw_id_fields	106
6.1.27	readonly_fields	107

- 6.1.28 save_as ·········· 107
- 6.1.29 save_as_continue ·········· 107
- 6.1.30 save_on_top ·········· 108
- 6.1.31 search_fields ·········· 108
- 6.1.32 show_full_result_count ·········· 110
- 6.1.33 sortable_by ·········· 110
- 6.1.34 view_on_site ·········· 110
- 6.1.35 自定义模板 ·········· 111
- 6.2 ModelAdmin 方法 ·········· 112
  - 6.2.1 save_model ·········· 112
  - 6.2.2 delete_model ·········· 113
  - 6.2.3 delete_queryset ·········· 113
  - 6.2.4 save_formset ·········· 113
  - 6.2.5 get_ordering ·········· 114
  - 6.2.6 get_search_results() ·········· 114
  - 6.2.7 save_related ·········· 114
  - 6.2.8 get_autocomplete_fields ·········· 115
  - 6.2.9 get_readonly_fields ·········· 115
  - 6.2.10 get_prepopulated_fields ·········· 115
  - 6.2.11 get_list_display ·········· 115
  - 6.2.12 get_list_display_links ·········· 115
  - 6.2.13 get_exclude ·········· 115
  - 6.2.14 get_fields ·········· 115
  - 6.2.15 get_fieldsets ·········· 115
  - 6.2.16 get_list_filter ·········· 115
  - 6.2.17 get_list_select_related ·········· 116
  - 6.2.18 get_search_fields ·········· 116
  - 6.2.19 get_sortable_by ·········· 116
  - 6.2.20 get_inline_instances ·········· 116
  - 6.2.21 get_inlines ·········· 116
  - 6.2.22 get_urls ·········· 116
  - 6.2.23 get_form ·········· 117
  - 6.2.24 get_formsets_with_inlines ·········· 117
  - 6.2.25 formfield_for_foreignKey ·········· 118
  - 6.2.26 formfield_for_manytomany ·········· 118
  - 6.2.27 formfield_for_choice_field ·········· 118
  - 6.2.28 get_changelist ·········· 119
  - 6.2.29 get_changelist_form ·········· 119
  - 6.2.30 get_changelist_formset ·········· 119
  - 6.2.31 lookup_allowed ·········· 119
  - 6.2.32 has_view_permission ·········· 120
  - 6.2.33 has_add_permission ·········· 120
  - 6.2.34 has_change_permission ·········· 120
  - 6.2.35 has_delete_permission ·········· 120
  - 6.2.36 has_module_permission ·········· 120
  - 6.2.37 get_queryset ·········· 120
  - 6.2.38 message_user ·········· 121
  - 6.2.39 get_paginator ·········· 121
  - 6.2.40 response_add ·········· 121
  - 6.2.41 response_change ·········· 121
  - 6.2.42 response_delete ·········· 121
  - 6.2.43 get_changeform_initial_data ·········· 122
  - 6.2.44 get_deleted_objects ·········· 122
  - 6.2.45 add_view ·········· 123
  - 6.2.46 change_view ·········· 123
  - 6.2.47 changelist_view ·········· 123
  - 6.2.48 delete_view ·········· 123
  - 6.2.49 history_view ·········· 123
- 6.3 ModelAdmin 资源 ·········· 123
- 6.4 定制验证功能 ·········· 124
- 6.5 InlineModelAdmin ·········· 124
  - 6.5.1 InlineModelAdmin.model ·········· 125
  - 6.5.2 InlineModelAdmin.fk_name ·········· 125
  - 6.5.3 InlineModelAdmin.formset ·········· 125
  - 6.5.4 InlineModelAdmin.form ·········· 125
  - 6.5.5 InlineModelAdmin.classes ·········· 125
  - 6.5.6 InlineModelAdmin.extra ·········· 125
  - 6.5.7 InlineModelAdmin.max_num ·········· 126
  - 6.5.8 InlineModelAdmin.min_num ·········· 126
  - 6.5.9 InlineModelAdmin.raw_id_fields ·········· 127
  - 6.5.10 InlineModelAdmin.template ·········· 128
  - 6.5.11 InlineModelAdmin.verbose_name ·········· 128
  - 6.5.12 InlineModelAdmin.verbose_name_plural ·········· 128
  - 6.5.13 InlineModelAdmin.can_delete ·········· 128
  - 6.5.14 InlineModelAdmin.show_change_link ·········· 129
  - 6.5.15 InlineModelAdmin.get_formset(request, obj=None, **kwargs) ·········· 129

6.5.16	InlineModelAdmin.get_extra(request, obj=None, **kwargs)·················· 129	
6.5.17	InlineModelAdmin.get_max_num (request, obj=None, **kwargs) ····· 129	
6.5.18	InlineModelAdmin.get_min_num (request, obj=None, **kwargs) ····· 130	
6.5.19	InlineModelAdmin.has_add_permission(request, obj)··············· 130	
6.5.20	InlineModelAdmin.has_change_permission(request, obj=None) ····· 130	
6.5.21	InlineModelAdmin.has_delete_permission(request, obj=None) ····· 130	
6.5.22	使用中间模型处理 ManyToMany 关系······························ 130	
6.6	重写管理后台模板······················ 131	
6.6.1	新建管理后台模板·············· 131	
6.6.2	重写与替换 ······················ 133	
6.6.3	可重写模板······················ 134	
6.6.4	根模板和登录模板·············· 134	
6.7	AdminSite ······························· 134	
6.7.1	重写 AdminSite ················ 134	
6.7.2	多管理后台的实现·············· 136	

第 7 章 路由系统······················ 137

7.1	Django 处理 HTTP 请求的流程······· 137
7.2	URLconf 示例···························· 137
7.3	URL 参数类型转化器··················· 138
7.4	自定义 URL 参数类型转化器········· 138
7.5	使用正则表达式 ························ 140
7.6	导入其他 URLconf ······················ 140
7.7	向视图传递额外参数··················· 141
7.8	动态生成 URL··························· 142
7.9	URL 名字和命名空间··················· 143

第 8 章 模型························· 145

8.1	模型简介································· 145
8.2	使用模型································· 145
8.3	字段····································· 146
8.3.1	AutoField ······················ 146

8.3.2	BigAutoField······················ 146	
8.3.3	BinaryField························ 146	
8.3.4	BooleanField ····················· 146	
8.3.5	CharField ························· 147	
8.3.6	DateField ························· 147	
8.3.7	DateTimeField ··················· 148	
8.3.8	DecimalField ····················· 148	
8.3.9	EmailField························ 149	
8.3.10	FileField··························· 149	
8.3.11	FilePathField····················· 150	
8.3.12	FloatField························· 151	
8.3.13	ImageField························ 151	
8.3.14	IntegerField······················ 151	
8.3.15	GenericIPAddressField········· 151	
8.3.16	PositiveIntegerField············ 152	
8.3.17	PositiveSmallIntegerField ····· 152	
8.3.18	SlugField·························· 152	
8.3.19	SmallIntegerField ··············· 152	
8.3.20	TextField ························· 152	
8.3.21	TimeField························· 152	
8.3.22	URLField·························· 152	
8.3.23	UUIDField························· 152	
8.4	字段参数·································· 153	
8.4.1	null································· 153	
8.4.2	blank······························ 153	
8.4.3	choices···························· 153	
8.4.4	default···························· 156	
8.4.5	help_text························· 156	
8.4.6	primary_Key····················· 156	
8.4.7	unique···························· 156	
8.4.8	verbose_name··················· 156	
8.5	表与表之间关系························· 157	
8.5.1	多对一关系······················ 157	
8.5.2	多对多关系······················ 157	
8.5.3	一对一关系······················ 158	
8.6	模型元属性······························ 158	
8.7	元属性··································· 158	
8.7.1	abstract··························· 159	
8.7.2	app_label························· 159	

8.7.3	base_manager_name	159
8.7.4	db_table	159
8.7.5	get_latest_by	159
8.7.6	order_with_respect_to	159
8.7.7	ordering	160
8.7.8	Indexes	160
8.7.9	constraints	161
8.7.10	verbose_name	161
8.7.11	verbose_name_plural	161
8.8	Manager 类	161
8.8.1	自定义 Manager 类	162
8.8.2	直接执行 SQL 语句	162
8.8.3	执行存储过程	165
8.9	数据增删改查	165
8.10	数据操作进阶——QuerySets	171
8.10.1	创建对象	172
8.10.2	修改对象	172
8.10.3	更新 ForeignKey	172
8.10.4	更新 ManyToManyField	173
8.10.5	数据查询	174
8.10.6	链式过滤器	179
8.10.7	查询条件	180
8.10.8	模型深度检索	183
8.10.9	多条件查询	184
8.10.10	主键查询	185
8.10.11	查询条件中的 % 和 _	185
8.10.12	F() 函数	186
8.10.13	Func() 表达式	187
8.10.14	QuerySet 和缓存	188
8.10.15	复杂查询与 Q 对象	189
8.10.16	模型比较	190
8.10.17	复制模型实例	190
8.10.18	批量更新	190
8.10.19	模型关系	191

第 9 章 视图 193

9.1	视图结构	193
9.2	HTTP 状态处理	193
9.3	快捷方式	195
9.3.1	render_to_string()	195
9.3.2	render()	195
9.3.3	redirect()	196
9.3.4	get_object_or_404()	197
9.3.5	get_list_or_404()	198
9.4	视图装饰器	198
9.4.1	HTTP 方法装饰器	199
9.4.2	GZip 压缩	199
9.4.3	Vary	200
9.4.4	缓存	201
9.5	Django 内置视图	202
9.5.1	serve	202
9.5.2	错误视图	203
9.6	HttpRequest 对象	204
9.6.1	属性	204
9.6.2	中间件属性	205
9.6.3	方法	206
9.6.4	QueryDict 对象	207
9.7	HttpResponse 对象	208
9.7.1	属性	209
9.7.2	方法	210
9.7.3	HttpResponse 子类	211
9.7.4	JsonResponse	211
9.7.5	FileResponse	212
9.8	TemplateResponse 对象	212
9.8.1	SimpleTemplateResponse 对象	212
9.8.2	TemplateResponse 对象	213
9.8.3	TemplateResponse 对象渲染过程	214
9.8.4	回调函数	215
9.8.5	使用 TemplateResponse 对象	215
9.9	文件上传	216
9.9.1	单一文件上传	216
9.9.2	多文件上传	218
9.9.3	临时文件	219
9.10	类视图	219
9.10.1	类视图入门	219
9.10.2	继承类视图	220
9.11	通用视图	220
9.11.1	通用视图	220

9.11.2 修改通用视图属性 ·················· 222
9.11.3 添加额外的上下文对象 ············ 222
9.11.4 queryset 属性 ························ 223
9.11.5 动态过滤 ······························ 224
9.11.6 通用视图与模型 ··················· 225
9.12 表单视图 ······························· 225
9.12.1 编辑表单视图 ······················ 226
9.12.2 当前用户 ···························· 228

第 10 章 模板 ································ 230
10.1 加载模板 ······························· 230
10.2 模板语言 ······························· 231
10.2.1 变量 ································· 231
10.2.2 标签 ································· 232
10.2.3 人性化语义标签 ··················· 236
10.2.4 过滤器 ······························ 239
10.2.5 注释 ································· 241
10.3 自定义标签和过滤器 ················ 241
10.3.1 编写自定义过滤器 ················ 242
10.3.2 编写自定义标签 ··················· 243
10.4 模板继承 ······························· 244

第 11 章 表单系统 ·························· 250
11.1 Form 类 ································· 250
11.2 表单字段类型 ························· 251
11.3 表单字段通用属性 ··················· 255
11.4 表单与模板 ···························· 255

第 12 章 中间件 ····························· 257
12.1 缓存中间件 ···························· 257
12.2 通用中间件 ···························· 258
12.3 GZip 中间件 ·························· 258
12.4 有条件的 GET 中间件 ·············· 258
12.5 语言环境的中间件 ··················· 258
12.6 消息中间件 ···························· 259
12.7 安全中间件 ···························· 259
12.8 会话中间件 ···························· 259
12.9 站点中间件 ···························· 259
12.10 身份验证中间件 ···················· 259

12.11 CSRF 保护中间件 ·················· 260
12.12 X-Frame-Options 中间件 ········ 260
12.13 中间件排序 ·························· 260
12.14 开发中间件 ·························· 261

第 13 章 自动化测试 ······················ 263
13.1 编写第一个测试用例 ················ 263
13.2 执行测试用例 ························· 264
13.3 修改代码中的 bug ··················· 264
13.4 边界值测试 ···························· 265
13.5 测试自定义视图 ······················ 266
13.6 测试 DetailView ······················ 268

第 14 章 国际化和本地化 ················ 270
14.1 名词解释 ······························· 270
14.2 翻译概述 ······························· 271
14.3 在 Python 中进行国际化 ··········· 271
14.3.1 注释 ································· 272
14.3.2 空操作 ······························ 273
14.3.3 复数 ································· 273
14.3.4 上下文标记 ························ 274
14.3.5 延迟翻译 ··························· 275
14.3.6 本地化的语言名 ··················· 276
14.4 编写模板代码 ························· 277
14.4.1 trans ································· 277
14.4.2 blocktrans ·························· 278
14.4.3 注释 ································· 279
14.5 翻译原理 ······························· 279

第 15 章 安全 ································ 282
15.1 网络攻击与保护 ······················ 282
15.1.1 跨站脚本攻击 ······················ 282
15.1.2 跨站请求伪造攻击 ················ 283
15.1.3 SQL 注入 ··························· 284
15.1.4 点击劫持 ··························· 284
15.2 检查配置信息 ························· 284

第 16 章 部署 ································ 286
16.1 WSGI 和 Application 对象 ········ 286

16.2　Ubuntu 部署 Django ············· 286
16.2.1　查看系统版本 ············· 286
16.2.2　更换国内源 ············· 286
16.2.3　查看 Python 版本 ············· 287
16.2.4　安装 pip3 ············· 288
16.2.5　安装 nginx ············· 289
16.2.6　安装 Django ············· 289
16.2.7　安装 uwsgi ············· 289
16.2.8　命令行运行网站 ············· 290
16.2.9　配置 uwsgi ············· 290
16.2.10　配置 nginx ············· 291
16.2.11　启动网站 ············· 291
16.2.12　设置静态文件 ············· 292
16.2.13　自启动服务 ············· 293

16.3　CentOS 部署 Django ············· 293
16.3.1　查看系统版本 ············· 293
16.3.2　更换国内源 ············· 294
16.3.3　更新 Python ············· 294
16.3.4　安装 Django ············· 296
16.3.5　安装 uwsgi ············· 296
16.3.6　命令行运行网站 ············· 296
16.3.7　配置 uwsgi ············· 296
16.3.8　安装 nginx ············· 297
16.3.9　自启动服务 ············· 298

附录 A　语言码 ············· **299**

附录 B　日期格式化字符串 ············· **301**

第1章
走进 Django 的世界

1.1 认识 Django

Django 是基于 Python 语言开发的一套重量级 Web 框架。设计初衷是帮助开发人员以最小的代码量快速建站。Django 通过丰富的内置功能帮助开发人员摆脱了以往 Web 开发中的很多困难，进而得以将更多精力专注于自己的网站开发中。

Django 基于 BSD 协议，完全开源，任何人都可以使用，Django 的 GitHub 地址是：https://github.com/django。

本书带领读者进入 Django 的世界，详细学习这套开发框架。

1.2 版本选择

从 1.0 版本开始，Django 按照 A.B 或 A.B.C 的形式命名版本编号。A.B 是主版本号，包含新功能以及对原有功能的改进，每一个新版本都向前兼容，Django 大概每 8 个月就会发布一个主版本；C 是小版本号，包含 bug 的修改等，每当有需要时就会发布。在 Django 正式版本发布之前，还会发布 alpha、beta 和 RC（Release Candidate，候选发布版本）版本，另外 Django 长期支持的版本用 LTS 表示。

Django 推荐使用最新的 Python 版本进行开发。Django 各个版本对 Python 的支持情况如下表所示（截至 2020 年 1 月）。

Django 版本	Python 版本
2.0	3.4，3.5，3.6，3.7
2.1	3.5，3.6，3.7
2.2	3.5，3.6，3.7，3.8（2.2.8 添加）
3.0	3.6，3.7，3.8

下图所示是 Django 官方对各个版本的支持情况。

Django 长期支持的版本包括 2.2、3.2 以及 4.2，当前的长期支持版本是 2.2，3.2 是下一个长期支持版本。

本书以 Django 3.0+Python 3.8 作为开发环境进行讲解。

为照顾大多数读者，本教程采用 Windows 系统作为开发环境。

1.3 搭建开发环境

1.3.1 安装 Python

本书采用最新的 Python 3.8 作为演示版本，读者可以根据个人情况选择其他版本，但是不要选择低于 3.6 的版本。具体安装方式不在本书介绍范围，读者可自行安装。

1.3.2 安装 Django

快速安装 Django 有 3 种方式。

方法 1：使用 pip 安装官方正式版本 Django。

这种方法可以快速安装最新的 Django 版本，如果没有特殊需求，强烈建议读者使用以下命令安装 Django：

```
$ pip install Django
```

方法 2：根据个人操作系统安装指定版本。

根据 Django 官方说法，目前有许多第三方分销商已经将 Django 集成到其软件包管理系统中，因此针对这些系统有不同的安装方式，如 CentOS 可以用以下命令安装 Django：

```
# yum install python-setuptools
# easy_install Django
```

方法 3：安装最新开发版本。

Django 的源代码托管在 GitHub，对于开发人员、乐于提前尝试新技术的人员或者 Django 的贡献作者，可以使用 git 客户端下载最新代码：

```
$ git clone https://github.com/django/django.git
```

下载并解压完 Django 源码，执行以下命令：

```
$ python -m pip install -e django/
```

通过以上命令可以将 Django 导入 Python 运行环境，使得代码可以方便地引用 Django 模块，同时该命令还安装了 django-admin。

> **注意**
>
> 如果已经安装过其他版本的 Django，可以先卸载旧版本再安装新版本，或者通过以下命令升级现有版本：
>
> ```
> $ pip install --upgrade django
> ```

安装结束，可以通过以下方式检查是否安装成功。

方式 1：通过如下命令。

```
$ pip list
```

结果如下图所示。

方式 2：输入如下命令。

```
python -m django -version
```

结果如下图所示。

方式 3：输入如下命令。

```
django.get_version()
```

结果如下图所示。

第 2 章
Web 开发框架

任何一门技术都有其知识体系，要想学好 Django，就必须先在头脑中对它的知识体系有一个清晰的认识。当了解了其知识体系中的每一个知识点时，即使不能够深刻理解它，也能够在遇到困难时迅速定位到问题的根源。

2.1 Socket 编程

Socket 也叫"套接字"，是计算机网络通信中最基础的内容，它通过对 TCP/IP 协议的封装提供了在不同主机之间进行通信的功能。当访问一个网站时，浏览器会打开一个套接字，通过套接字建立与服务器之间的链接，链接建立成功后服务器提供对访问的响应并返回访问内容，浏览器接收响应并显示。

几乎所有 Web 应用都是通过 Socket 实现的。一个网站本质上就是 Socket 服务端和客户端之间的通信，Web 服务器就是服务端，用户浏览器就是客户端。用户访问网站的过程就是服务端与客户端 Socket 通信的过程，如下图所示。

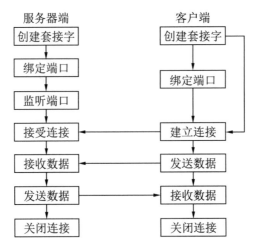

下面的程序是一个简单的 Socket Web 服务器。运行程序，通过浏览器访问 http://127.0.0.1:8000/，就会打开一个"Hello, World!"的页面。

```
#!/usr/bin/env python
# -*- coding: UTF-8 -*-
```

```python
import socket

def handle_request(client):
    buf = client.recv(1024)
    client.send(bytes("HTTP/1.1 200 OK\r\n\r\n".encode('utf_8')))
    client.send(bytes("Hello, World!".encode('utf_8')))

def main():
    sock = socket.socket(socket.AF_INET, socket.SOCK_STREAM)
    sock.bind(('localhost',8000))
    sock.listen(5)

    while True:
        connection, address = sock.accept()
        handle_request(connection)
        connection.close()

if __name__ =='__main__':
    main()
```

浏览器访问效果如下图所示。

这就是所有网站的实现原理：接收 HTTP 请求，解析 HTTP 请求，发送 HTTP 响应。如果这些工作都由网站开发人员来做，那么开发人员不仅需要熟悉自身产品相关的技术，而且需要学习 HTTP 协议、TCP/IP 协议等协议，这会带来很多额外的工作量。幸运的是，这些工作已经有人帮我们完成了，在 Python 中这个工作由 WSGI 接口实现，而 Django 是基于 WSGI 接口的。

当访问同一个网站时，如果输入的 URL 不同，网页显示的内容也不同，这就是一般 Web 框架所实现的。接下来开发一个可以根据用户输入 URL 的不同而显示不同页面信息的 Web 框架。这个框架暂时可以接收两个地址：index 和 detail，如果输入其他地址则返回 404 错误，具体代码如下：

```python
#!/usr/bin/env python
# -*- coding: UTF-8 -*-

from wsgiref.util import setup_testing_defaults
from wsgiref.simple_server import make_server

def simple_app(environ, start_response):
    setup_testing_defaults(environ)

    status = '200 OK'
    headers = [('Content-type', 'text/html; charset=utf-8')]
```

```
        start_response(status, headers)

        url = environ["PATH_INFO"]

        response = ''
        if url == '/index':
            response = '<h1>这里是 Index 页面</h1>'
        elif url == '/detail':
            response = '<h1>这里是 Detail 页面</h1>'
        else:
            response = '<h1 style="color:red;">网页丢失了：404</h1>'

        return [response.encode("utf-8")]

if __name__ == '__main__':
    httpd = make_server('', 8000, simple_app)
    httpd.serve_forever()
```

执行脚本，然后分别访问 index、detail 和 home 页面（注意，在代码中并没有处理 home 请求），浏览器显示效果如下面三幅图所示。

可以看到，由于在这个 Web 框架中处理了 index 请求（if url == '/index':）和 detail 请求（elif url == '/detail'），因此访问这两个页面时网页能够正常显示；对于其他请求统一按照 404 的方式进行处理，访问 home 页面时抛出 404 页面。

虽然这个 Web 框架看起来非常简单，但是事实上很多开发框架都是这样在 WSGI 基础上开发的，只是不同的框架提供了不同的功能而已。如果你感兴趣，可以尝试自己开发一个 Web 框架。

2.2 MTV 框架

前面的 Web 开发框架非常简单，完全符合初级程序员想到哪做到哪的开发风格。这种开发风格在个人开发或者微型团队开发中不会出现任何问题，有时还能够提高沟通效率，减少工作量。但是，当团队规模扩大、业务场景变得越来越复杂的情况下，这种开发模式就会给开发人员和技术管理人员带来诸多麻烦，例如，有时某一个开发人员辛辛苦苦地完成一个方法后却发现其他人在很早之前已经完成了；有时系统里存在一个非常高效的方法却没有人知道存放位置，最终的代码变得杂乱无章且难以管理。为了解决这些问题，软件开发中逐渐引入了开发框架的概念，开发框架通常针对某一领域使得代码更容易地被重用。经常被提及的设计模式有微软的 ASP.NET MVC 框架、Java 的 Spring 框架等。

Django 框架的基础是 MTV 模式，它将开发任务分为三大部分：Model、Template、View。很多人可能对 MTV 不太了解，但是，如果说到一个和它相似的开发模式你一定了解，或者说很熟悉，那就是 MVC 开发模式。MVC 模式就是把 Web 应用分为 Model（模型）、View（视图）、Controller（控制器）三层。

- Model：负责业务对象与数据库的关系映射（一般基于 ORM（Object Relational Mapping，对象关系映射）框架）。
- View：负责页面展示，也就是与用户直接交互的网页部分。
- Controller：接收并处理用户的请求，通常需要调用 Model 和 View 来完成用户请求。

MVC 模式三者之间的关系如下图所示。

MTV 与 MVC 模式非常相似，也将开发工作分为三层：

- M 代表模型（Model）：负责业务对象和数据库的关系映射（ORM），这与 MVC 模式中的模型是一样的。
- T 代表模板（Template）：负责把页面展示给用户（html），这部分类似于 MVC 中的视图。
- V 代表视图（View）：负责业务逻辑，并在适当时候调用 Model 和 Template，这里就不是 MVC 的 View 了，反而更像是 Controller。

Django 的响应模式如下图所示。

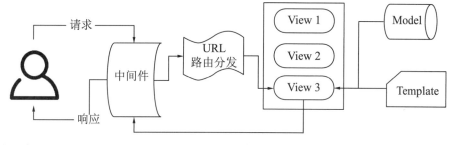

响应顺序如下：

（1）Django 中间件接收到一个用户请求。
（2）Django 通过 URLconf 查找对应的视图，然后进行 URL 路由分发。
（3）视图接收请求，查询对应的模型，调用模板生成 HTML 文档。
（4）视图返回处理后的 HTML 文档。
（5）Web 服务器将响应内容发送给客户端。

第 3 章
搭建第一个 Django 网站

学习任何新技术都不是一件容易的事情,很多开发人员喜欢花费大量时间进行碎片化学习,这种学习方式虽然能够满足一时的工作需要,但是并不能帮助深入、全面理解一门技术,因此很多开发人员在使用一门技术很长时间之后还是不能灵活应用它。因此,本书首先带领读者使用 Django 框架快速搭建一个投票网站,使读者能够从整体上认识 Django,然后详细介绍技术细节,最后帮助读者深入理解并掌握 Django。

本章所演示的投票网站主要包含以下两部分:
- 公开的网站前台部分,用于浏览民意测验结果以及进行网上投票。
- 网站后台管理功能,允许管理员添加、修改、删除调查问卷。

3.1 创建项目

首先新建一个名为 django3 的文件夹,打开命令行提示符,在命令行提示符窗口输入"cd django3 文件夹的路径",按回车键将命令行切换到 django3 文件夹,然后输入下面命令:

```
> django-admin startproject mysite
```

命令执行结束后,将会在 django3 文件夹下创建一个名为 mysite 的文件夹。

注意

- 应避免使用 Python 内置的包或者 Django 内嵌组件名来命名项目,例如不能使用 django 来命名新项目,因为这会与 Django 自身产生冲突,也不能使用 test 作为项目名,因为这会与 Python 的内置包产生冲突。
- 不要将 Django 项目代码文件与其他网站项目放在一起,例如不应将 Django 文件放置在 Web 服务器的根目录,因为这样可能会将 Django 的代码暴露在浏览器中。

此时 mysite 文件夹下的文件目录结构如下:

```
mysite/
    manage.py
    mysite/
        __init__.py
        settings.py
        urls.py
```

```
            asgi.py
            wsgi.py
```

以上目录和文件的意义如下：

- 最外层文件夹"mysite"是整个项目的容器，它的名字对于 Django 来说没有任何意义，虽然创建项目的时候使用了 mysite 作为项目名字，但是随时可以对它进行重命名。
- manage.py 脚本文件是一个命令行工具，通过使用这个文件可以管理 Django 项目，后面章节会对 django-admin 和 manage.py 进行详细介绍。
- 第二级的"mysite"文件夹才是当前 Django 工程所使用的 Python 包（包含 __init__.py 文件的 Python 文件夹）。这个文件夹的名字将会被用来导入包内的所有内容（例如导入 mysite.urls）。
- mysite/__init__.py：表明当前文件夹是一个 Python 包。
- mysite/settings.py：当前 Django 工程的配置文件，后续章节会对 Django 的配置进行详细介绍。
- mysite/urls.py：当前 Django 工程的路由配置文件，包含工程的路由信息，后面章节会对 Django 的路由系统进行详细介绍。
- mysite/asgi.py：Django 3.0 新增对 ASGI 的支持，这是对 WSGI 的一个补充。ASGI（Asynchronous Server Gateway Interface）为可异步的 Python Web 服务、框架和应用提供标准接口，使得项目可以很好地支持 HTTP、HTTP2、WebSocket 等协议。
- mysite/wsgi.py：兼容 WSGI 的 Web 服务入口。Django 应用程序是基于 WSGI 服务开发的，因此运行或部署 Django 程序时需要指定 WSGI 配置信息，在后面章节中会介绍如何使用 WSGI 部署 Django 应用程序。

3.2 运行项目

到目前为止，我们已经搭建好一个最简单的 Django 工程，下面来检查这个工程是否能够正常运行。将命令行提示符所在位置切换到内部的 mysite 文件夹（与 manage.py 同级），执行以下命令：

```
> python manage.py runserver
```

运行结果如下：

```
Watching for file changes with StatReloader
Performing system checks...

System check identified no issues (0 silenced).

You have 17 unapplied migration(s). Your project may not work properly until you
apply the migrations for app(s): admin, auth, contenttypes, sessions.
Run 'python manage.py migrate' to apply them.
January 03, 2020 - 10:42:30
Django version 3.0, using settings 'mysite.settings'
Starting development server at http://127.0.0.1:8000/
Quit the server with CTRL-BREAK.
```

暂时忽略上面输出结果中的警告信息"You have 17 unapplied migration(s)"。这是因为新建的 Django 工程还没有同步数据库信息。

此时我们已经使用一个 Python 内嵌的轻量级 Web 服务器运行了 Django 工程。这也是 Django 能够快速开发 Web 应用程序的一个优势——在开发过程中不需要关心 Web 服务。

如果细心观察会发现在 mysite 文件夹的同级多出一个 db.sqlite3 数据库文件。

现在 Django 应用已经运行起来，打开浏览器在地址栏输入 http://127.0.0.1:8000/，此时能看到如下图所示的 Django 欢迎页面，说明 Django 程序已经创建成功。

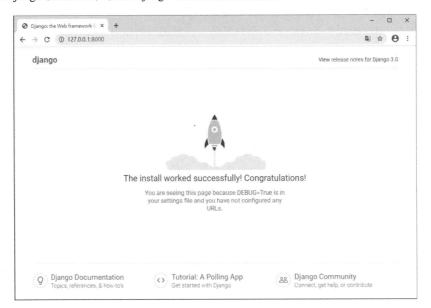

> **注意**
>
> 这种运行 Django 应用程序的方式在稳定性和网站性能方面都很差，只适用于开发过程，绝对不能应用在生产环境，如果使用这种方式部署 Django 网站，当用户登出服务器时或者远程会话终止时，Web 服务也会停止。

对于 runserver 命令，第 4 章中会详细介绍。

3.3 创建投票应用

前面已经完成了 Django 工程的创建，接下来开始创建应用程序。每一个 Django 应用程序都是一个 Python 包。django-admin 和 manage.py 可以帮助开发人员快速创建应用程序文件夹，因此大大地提高了开发效率。

> **项目（project）与应用程序（APP）**
>
> 前面多次提到 Django 项目与应用程序，那么项目与应用程序之间有什么区别呢？其实应用程序是真正工作的组件，例如一个博客系统或者投票系统。项目是包含网站配置信息和应用程序等的集合，一个项目可以包含多个应用程序，而一个应用程序也可以属于多个项目。
>
> 应用程序可以放置在任何 Python 路径能够识别的地方，在本书中，将应用程序放在 manage.py 的同级目录，这样方便调用。

切换到 manage.py 所在目录,然后执行以下命令:

```
> python manage.py startapp polls
```

命令执行结束就会在 mysite 同级目录创建应用程序 polls,polls 的目录结构如下:

```
polls/
    __init__.py
    admin.py
    apps.py
    migrations/
        __init__.py
    models.py
    tests.py
    views.py
```

3.4 开发第一个视图

Django 的视图是负责页面展示的重要模块,用于处理网站业务逻辑。

打开 polls/view.py 文件,添加以下代码:

```python
#!/usr/bin/python
# -*- coding: UTF-8 -*-

from django.http import HttpResponse

def index(request):
    return HttpResponse("你好!这里是在线投票系统。")
```

一个最简单的 Django 视图已经创建完成,为了能够访问它,需要在 URL 中添加路由映射。在 polls 文件夹下创建文件 urls.py,并在 urls.py 文件中添加以下代码:

```python
#!/usr/bin/python
# -*- coding: UTF-8 -*-

from django.urls import path

from . import views

urlpatterns = [
    path('', views.index, name='index'),
]
```

接下来需要在 mysite/urls.py 中引用 polls/urls.py,修改 mysite/urls.py 如下:

```python
from django.contrib import admin
from django.urls import include, path

urlpatterns = [
    path('polls/', include('polls.urls')),
    path('admin/', admin.site.urls),
]
```

函数 include() 可以用来引用其他 URLconfs(urls.py)。通过合理使用 include() 函数可以将整个网站中的所有 URL 分配到多个文件中,使代码更加简洁合理。

> **注意**
> 除了 admin.site.urls 之外，在任何时候都应该使用 include() 函数引用其他路由模块。

到目前为止，Django 项目中已经包含了一个视图。重新调用 runserver 命令启动 Web 服务，查看该视图是否能够正常工作。

在浏览器中输入 http://127.0.0.1:8000/polls/，按回车键，显示效果如下图所示。

关于 path() 函数的详细用法请参考第 7 章。

3.5 配置数据库

前面提到 Django 应用程序的配置信息都存储在 mysite/settings.py 文件中，数据库配置也不例外。settings.py 是一个标准的 Python 模块，其中存放了很多模块变量，数据库配置信息就是其中的一个变量。默认情况下 Django 使用 SQLite 作为数据库。SQLite 是一个免安装的数据库系统，非常简单易学，Python 已经提供了相应的支持模块，因此不需要做任何事情就可以在 Django 中使用 SQLite。

虽然 SQLite 有如此多的优势，但是，当你将 Django 程序真正应用到生产环境时，可能还是会因为各种问题而不得不更换数据库，事实上几乎没有人在生产环境使用 SQLite。因此，Django 官方提供了对 4 种数据库的支持：PostgreSQL、MySQL、Oracle 和 SQLite。对于不同的数据库，Django 提供了不同的数据库绑定（database binding），对此后续会详细介绍，本章使用默认参数就可以了。

下面是默认的数据库配置：

```
DATABASES = {
    'default': {
        'ENGINE': 'django.db.backends.sqlite3',
        'NAME': os.path.join(BASE_DIR, 'db.sqlite3'),
    }
}
```

关于更多的数据库配置信息请参考第 5 章。

数据库配置完成后就可以迁移（migrate）数据库了，这也一并解决了前面运行 runserver 命令时的异常"You have 17 unapplied migration(s)"：

```
> python manage.py migrate
```

migrate 命令根据 settings.py 中的 INSTALLED_APPS 创建必要的数据库表，每一个 Django 项目都会默认启用一些应用，举例如下。

- django.contrib.admin：管理员站点，你很快就会使用它。
- django.contrib.auth：认证授权系统。
- django.contrib.contenttypes：内容类型框架。

- django.contrib.sessions：会话框架。
- django.contrib.messages：消息框架。
- django.contrib.staticfiles：管理静态文件的框架。

命令行会显示出执行了哪些脚本，从脚本名字能够大致推断出创建了哪些数据库表。也可以使用数据库客户端程序打开数据库，在这里使用 SQLiteStudio 管理我的数据库，如下图所示。

此时可以看到数据库已经包含 10 个表，而某些表中还同时添加了数据，例如 auth_permission 表，如下图所示。

id	content_type_id	codename	name
1	1	add_logentry	Can add log entry
2	1	change_logentry	Can change log entry
3	1	delete_logentry	Can delete log entry
4	1	view_logentry	Can view log entry
5	2	add_permission	Can add permission
6	2	change_permission	Can change permission
7	2	delete_permission	Can delete permission
8	2	view_permission	Can view permission
9	3	add_group	Can add group
10	3	change_group	Can change group
11	3	delete_group	Can delete group
12	3	view_group	Can view group
13	4	add_user	Can add user
14	4	change_user	Can change user
15	4	delete_user	Can delete user
16	4	view_user	Can view user
17	5	add_contenttype	Can add content type
18	5	change_contenttype	Can change content type
19	5	delete_contenttype	Can delete content type
20	5	view_contenttype	Can view content type
21	6	add_session	Can add session
22	6	change_session	Can change session
23	6	delete_session	Can delete session
24	6	view_session	Can view session

以上就是 migrate 命令为我们做的所有事情，关于 migrate 命令的更多用法请参考第 4 章。

3.6 创建模型

事实上我们绝不能满足于现有的数据库结构，不同的应用需要不同的数据库表，我们需要学会创建自己的模型。

现在开始创建模型（model），在详细学习 Django 的 ORM 开发之前，读者只要将模型理解为数据库表对应的 Python 类的表现形式即可。每一个模型对应一个数据库表，而模型的属性就是数据库表的字段。

在线投票系统需要两个模型：问卷（Question）和选项（Choice）。Question 包含两个字段 question_text（问卷描述）和 pub_date（问卷发布时间）；Choice 包含两个字段 choice_text（选项内容）

和 votes（选项得票数），另外由于每一个选项都必须属于一个问卷，因此需要给选项一个问卷外键。结合以上分析，修改 polls/models.py 文件，完成后的模型代码如下：

```python
#!/usr/bin/python
# -*- coding: UTF-8 -*-

from django.db import models

class Question(models.Model):
    question_text = models.CharField(max_length=200)
    pub_date = models.DateTimeField('date published')


class Choice(models.Model):
    question = models.ForeignKey(Question, on_delete=models.CASCADE)
    choice_text = models.CharField(max_length=200)
    votes = models.IntegerField(default=0)
```

每一个模型类都是 django.db.models.Model 的子类，而模型的每一个属性都是 Field 类的实例，表示一个数据库表的字段。

每个 Field 类实例变量的名字都是数据库字段名（例如 question_text 和 pub_date），因此在给字段起名字的时候一定要注意是否适合数据库。

为了满足不同数据库的需要，Django 提供了几十个 Field 子类，不同的 Field 类在实例化的时候会接收不同的参数，这些会在后续内容中详细介绍。

3.7 激活模型

前面提到所有已经启用的 Django 应用程序都会被记录在 INSTALLED_APPS 中，而 migrate 命令也需要去 INSTALLED_APPS 检索所有应用程序，所以为了将前面创建的模型写入数据库，还需要激活它，激活应用程序的方式就是把应用程序的配置文件加入 INSTALLED_APPS 中。

Polls 项目的配置文件存放在 polls/apps.py 脚本中，默认的类名字是 PollsConfig，因此对应的 Python 路径就是"polls.apps.PollsConfig"：

```python
INSTALLED_APPS = [
    'django.contrib.admin',
    'django.contrib.auth',
    'django.contrib.contenttypes',
    'django.contrib.sessions',
    'django.contrib.messages',
    'django.contrib.staticfiles',
    'polls.apps.PollsConfig',
]
```

项目激活之后，执行以下命令生成数据库表：

```
> python manage.py makemigrations polls
```

makemigrations 命令会检测模型文件的修改，并且把修改的部分存储成一次迁移（migrate），这个迁移就是一个脚本文件，默认会保存在 migrations 文件夹下。这里不需要关心这些迁移文件具体做了什么，更不需要了解它是怎么做的，唯一要做的就是使用 migrate 命令将这些迁移应用到数据库中：

```
> python manage.py migrate
```

这个 migrate 命令选中所有还没有执行过的迁移（Django 通过在数据库中创建一个特殊的表 django_migrations 来跟踪执行过哪些迁移）并应用在数据库上，也就是将模型的更改同步到数据库结构上。

迁移命令非常强大，可以使不懂数据库开发的开发人员很轻松地操作数据库，后续内容会进一步介绍 migrate 命令。

提示

将数据库更新拆分成 makemigrations 和 migrate 两个命令可以方便地使用源代码管理工具管理数据库的变更记录。

3.8 Django 管理页面

Django 的强大体现在其内置的 Admin 模块可以使得开发人员在不做任何编码的情况下就拥有网站后台管理功能。

执行以下命令创建网站超级管理员：

```
> python manage.py createsuperuser
```

按照命令提示依次输入用户名、邮箱地址、密码并重复输入密码。

超级管理员创建成功后，重新启动 Web 服务：

```
>python manage.py runserver
```

在地址栏输入网址：http://127.0.0.1:8000/admin/，打开后 Django 后台登录页面如下图所示。

在登录页面输入刚刚创建的超级管理员用户名和密码，单击"登录"按钮进入后台管理页面，如下图所示。

从 Django 2.0 开始，Django 管理后台对不同尺寸的浏览器做了自适应，这对于开发人员来说真是一个里程碑式的改进，用户不用再通过左右拖动来查看网页内容了。

下图是一个小尺寸 PC 浏览器的显示样式。

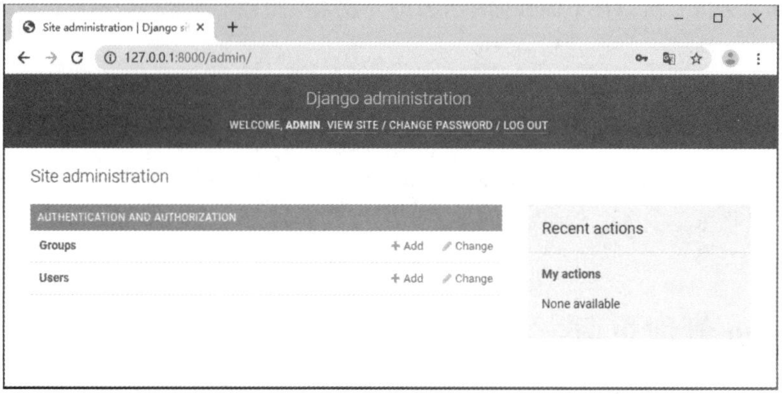

在手机上也能够很好地打开 Django 管理后台，下图是在 iPhone X 上的显示效果。

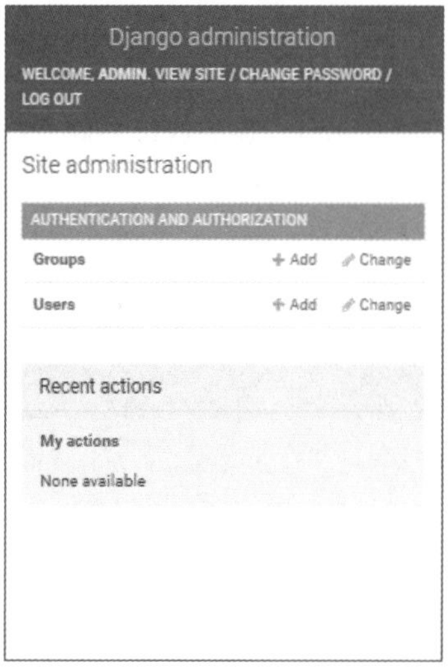

3.9 向管理页面中添加投票应用

到目前为止，投票系统已经有了超级管理员账号，也有了网站后台管理系统，但是，从前面的截图可以看出，后台系统还缺少对投票应用基本数据的修改功能，如没有问卷发布功能。接下来看看如何使得 Django 管理后台能够添加并修改问卷。

打开 polls/admin.py 文件，添加以下代码：

```
from django.contrib import admin
from .models import Question

admin.site.register(Question)
```

重启 Web 服务并刷新后台管理页面，如下图所示，可以看到此时多了一个 POLLS 节点。

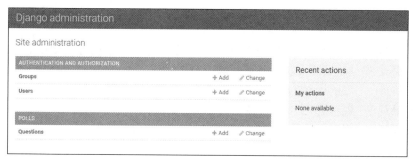

在 POLLS 节点下有一行 Questions。Questions 是一个超链接，单击它可以查看全部已有问卷。由于目前系统中还不存在任何问卷，单击 Add 按钮添加一条问卷信息，如下图所示。

点击 SAVE 按钮保存问卷，保存之后网页自动跳转到问卷列表页面，可以看到问卷列表下多了一条记录，如下图所示。

单击 Question object (1) 进入问卷编辑页面，如下图所示。

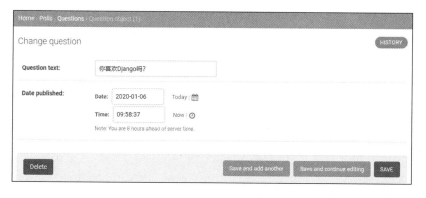

3.10 添加视图

现在网站的后台管理模块已经可以工作了，还缺少前台页面。投票系统需要以下几个页面：
- 问题索引页——展示最近的几个投票问题。
- 问题详情页——展示某个投票的问题和不带结果的选项列表。
- 问题结果页——展示某个投票的结果。
- 投票处理器——用于响应用户为某个问题的特定选项投票的操作。

在 Django 中每一个页面或者其他内容都是通过视图呈现出来的，每一个视图就是一个 Python 函数或者类方法，Django 中的视图是"一类具有相同功能和模板的网页的集合"。Django 通过 URL 确定调用哪一个视图，Django 的 URL 相较于早期网站的 URL 更加简洁优雅。

Django 通过 URLconfs 将 URL 模式字符串与视图关联起来，URL 模式字符串就是一个 URL 的一般形式，如 /newsarchive/<year>/<month>/。

在 polls/views.py 文件中添加以下视图：

```python
def detail(request, question_id):
    return HttpResponse("将为您打开问卷 %s。" % question_id)

def results(request, question_id):
    response = "正在查看问卷 %s 的结果。"
    return HttpResponse(response % question_id)

def vote(request, question_id):
    return HttpResponse("请为问卷 %s 提交您的答案。" % question_id)
```

修改 polls.urls 文件，添加以下 URL 映射：

```python
urlpatterns = [
    path('', views.index, name='index'),
    path('<int:question_id>/', views.detail, name='detail'),
    path('<int:question_id>/results/', views.results, name='results'),
    path('<int:question_id>/vote/', views.vote, name='vote'),
]
```

好了，重启 Web 服务器，在浏览器中访问 http://127.0.0.1:8000/polls/24/，如下图所示。

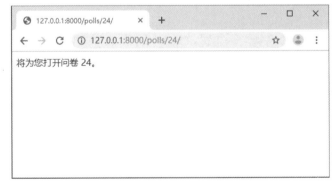

继续访问 http://127.0.0.1:8000/polls/24/results/ 和 http://127.0.0.1:8000/polls/24/vote/，同样能够正常显示视图内容，如下面两幅图所示。

之所以 Django 能够正常调用解析 URL，是因为在 settings.py 中设置了 ROOT_URLCONF = 'mysite.urls'。当用户访问的 URL 包含 polls/ 时，Django 会根据 mysite.urls 中的设置，跳转到 polls.urls 并进行验证，直到找到第一个匹配的 URL 为止。

以上视图中参数 question_id 的值来自于 <int:question_id>。<int:question_id> 用于匹配 URL 中的值，并将捕捉到的值作为关键字参数传递给视图，其中 :question_id 对应视图的参数，int: 决定了 URL 中的哪类值符合匹配条件。

3.10.1 扩展视图

每一个视图都应该负责一个具体的业务逻辑，视图执行结束会返回一个包含页面内容的 HttpResponse 对象或者异常信息。

下面修改 index 视图使它返回最新的 5 条调查问卷。

```
from .models import Question

def index(request):
    latest_question_list = Question.objects.order_by('-pub_date')[:5]
    output = ', '.join([q.question_text for q in latest_question_list])
    return HttpResponse(output)
```

代码 Question.objects.order_by('-pub_date') 是 Django 的数据库 API 语法，用于从数据库中查找数据，在介绍模型时将进行详细讲解。

访问 index 页面以查看显示情况，如下图所示。

此时调查问卷已经显示到网页上，但是可以发现在 index 视图中使用了硬编码，如果想要修改网页显示样式就需要重新编写 Python 代码。对此 Django 提供了一套模板系统（templates），可以将业务逻辑与页面显示样式分离开。下面来看看如何使用模板系统。

首先在 polls 文件夹下创建一个新文件夹 templates，为了目录结构清晰，在 templates 文件夹下再创建一个 polls 文件夹，最后在 polls 下创建一个 index.html 文件。这个 index.html 就是即将应用于 index 视图的模板。

在 settings.py 中有一个关于模板的配置项：TEMPLATES。Django 就是根据这个配置查找并解析模板的，具体工作原理会在第 5 章进行讲解。

将下面代码写入模板文件 index.html：

```
{% if latest_question_list %}
    <ul>
    {% for question in latest_question_list %}
        <li><a href="/polls/{{ question.id }}/">{{ question.question_text }}</a></li>
    {% endfor %}
    </ul>
{% else %}
    <p>还没有调查问卷！</p>
{% endif %}
```

接下来修改 index 视图：

```
from django.http import HttpResponse
from django.template import loader

from .models import Question

def index(request):
    latest_question_list = Question.objects.order_by('-pub_date')[:5]
    template = loader.get_template('polls/index.html')
    context = {
        'latest_question_list': latest_question_list,
    }
    return HttpResponse(template.render(context, request))
```

新视图会从模板文件夹下加载模板文件并将一个字典对象传入视图。

重启 Web 服务器，刷新 index 页面，效果如下图所示。

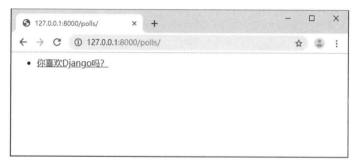

上面代码的工作原理是先使用 loader 方法加载模板文件并向它传递一个上下文对象（context），然后使用 HttpResponse 方法初始化一个 HttpResponse 对象并返回给浏览器。由于很多 Django 视图都是这样工作的，因此 Django 提供了一个简写函数：render()。下面使用 render 函数重写 index 视图：

```python
from django.shortcuts import render
from .models import Question

def index(request):
    latest_question_list = Question.objects.order_by('-pub_date')[:5]
    context = {'latest_question_list': latest_question_list}
    return render(request, 'polls/index.html', context)
```

此时重新访问 index 页面时可以发现效果与之前一样，但是不再需要 loader 和 HttpResponse 方法。

3.10.2 处理 404 错误

404 错误是一个比较常见的网页访问错误，当被访问的 URL 资源不存在时就会抛出这类错误。下面修改 detail 视图使其在无法查找到问卷的时候抛出 404 错误。

```python
from django.http import Http404
from django.shortcuts import render
from .models import Question

def detail(request, question_id):
    try:
        question = Question.objects.get(pk=question_id)
    except Question.DoesNotExist:
        raise Http404("问卷不存在")
    return render(request, 'polls/detail.html', {'question': question})
```

按照前面步骤在 polls 文件夹下创建一个 detail.html 文件作为 detail 视图的模板文件，模板内容暂时用 {{ question }} 表示。

此时重启 Web 服务，分别访问一个存在的和一个不存在的问卷，效果如下面两幅图所示。

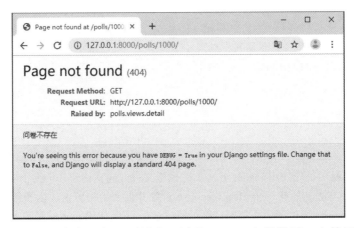

由于 404 错误是一个非常常见的网页异常，因此 Django 也提供了一个简写方法：get_object_or_404。下面使用 get_object_or_404() 修改 detail 视图：

```
from django.shortcuts import get_object_or_404, render

def detail(request, question_id):
    question = get_object_or_404(Question, pk=question_id)
    return render(request, 'polls/detail.html', {'question': question})
```

重新访问 detail 页面，效果如下图所示。

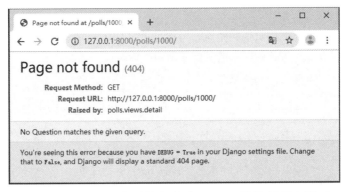

此时网页仍然抛出 404 错误，不过错误信息变成 Django 默认的英文形式，此时可以通过修改 get_object_or_404() 方法源代码的方式修改错误信息，修改完需要重启 Web 服务。

与 get_object_or_404 相似，Django 还提供了一个判断 list 是否存在的方法：get_list_or_404，在此不做详细介绍。

3.11 使用模板系统

前面的 detail.html 模板过于简单，现实中 Django 的模板系统非常强大，可以制作丰富多彩的网页效果。将以下代码复制到模板文件 detail.html 中：

```
<h1>{{ question.question_text }}</h1>
<ul>
{% for choice in question.choice_set.all %}
```

```
        <li>{{ choice.choice_text }}</li>
    {% endfor %}
</ul>
```

上面代码中的双大括号形式（{{ }}）是 Django 模板语言中的属性访问语法，采用英文句点的方式访问变量的属性，如示例中的代码 {{ question.question_text }}，其中 question 是视图通过字典形式传递给模板的变量，通过"."访问 question 的属性。

模板中 {% %} 形式的代码是 Django 模板语言的函数语法，上例中 {% for choice in question.choice_set.all %} 是一个 for 循环，循环对象是 question.choice_set.all，该对象等价于 Python 语法中的 question.choice_set.all()，返回一个可迭代的数组。Django 模板函数需要有结束标记，本例中 {% for %} 循环的结束标记是 {% endfor %}。

3.11.1 模板中的超链接

在前面 polls/index.html 模板中，使用硬编码的形式编写 HTML 超链接：

```
<li><a href="/polls/{{ question.id }}/">{{ question.question_text }}</a></li>
```

当项目中存在很多模板并且多个模板都使用同一个 URL 的时候，如果需要修改 URL，那么这种书写方式会给开发人员带来很大的工作量。此时可以通过对 URL 命名的方式解决这类问题，之前介绍 URL 时讲到了 URL 的命名，本例中的 URL 如下：

```
path('<int:question_id>/', views.detail, name='detail'),
```

使用 URL 重新修改模板如下：

```
<li><a href="{% url 'detail' question.id %}">{{ question.question_text }}</a></li>
```

其中，{% url %} 是 Django 的模板标签，用于定义 URL。该标签将会在 polls/urls 模块中查找名为"detail"的 URL，question.id 作为参数传递给 URL，如果需要传递多个参数时，只要在 question.id 后面紧跟一个空格然后继续添加参数即可。

通过 {% url %} 模板标签可以快速修改模板中的 URL，极大地提高工作效率，以保证代码安全。例如在这个 URL 中增加一个节点，如改成 polls/specifics/12/，此时只要修改 urls.py 中的定义即可：

```
path('specifics/<int:question_id>/', views.detail, name='detail'),
```

3.11.2 为超链接添加命名空间

命名空间可以有效地隔离变量，防止出现名称相同的变量之间调用混乱的问题。Django 中可以为 URL 定义命名空间。试想一下，在真实项目中往往会存在很多应用程序，而不同应用程序之间可能存在同名的视图，如多个应用程序中都存在 detail 视图，那么在 {% url %} 标签中如何确定调用哪一个应用程序中的 URL 呢？此时可以通过为 URL 添加命名空间的方式解决以上问题。

打开 polls/urls.py 文件，在其中添加 app_name 变量来设置 URLconf 的命名空间，修改后的代码如下：

```
#!/usr/bin/python
# -*- coding: UTF-8 -*-

from django.urls import path
```

```python
from . import views

app_name = 'polls'
urlpatterns = [
    path('', views.index, name='index'),
    path('<int:question_id>/', views.detail, name='detail'),
    path('<int:question_id>/results/', views.results, name='results'),
    path('<int:question_id>/vote/', views.vote, name='vote'),
]
```

接下来修改 polls/index.html 模板中的 URL，为 detail 视图添加命名空间：

```
<li><a href="{% url 'polls:detail' question.id %}">{{ question.question_text }}</a></li>
```

此时单击 index 页面中的超链接仍能正常显示。

3.12　HTML 表单

HTML 表单是客户端与服务器端进行交互的重要方式，在讲解 HTML 表单之前先进行必要的准备工作。

步骤 1：将 Choice 模型注册到 admin 网站，修改 polls/admin.py。

```python
from django.contrib import admin
from .models import Question, Choice

admin.site.register(Question)
admin.site.register(Choice)
```

步骤 2：登录 admin 后台，为问卷添加选项。

单击 Add 按钮，进入添加问卷选项页面，效果如下图所示。

添加 3 个选项，效果如下面三幅图所示。

完成准备工作后，继续修改 polls/detail.html 模板，为其添加 HTML 表单用于提交信息，新模板如下：

```
<h1>{{ question.question_text }}</h1>

{% if error_message %}<p><strong>{{ error_message }}</strong></p>{% endif %}

<form action= "{% url 'polls:vote' question.id %}" method= "post" >
{% csrf_token %}
{% for choice in question.choice_set.all %}
    <input type= "radio" name= "choice" id= "choice{{ forloop.counter }}" value= "{{ choice.id }}" >
    <label for="choice{{ forloop.counter }}">{{ choice.choice_text }}</label><br>
{% endfor %}
    <input type="submit" value="Vote">
</form>
```

代码分析：
- 在问卷详情页显示问卷相关选项，并为每一个选项添加单选按钮（radio）。
- 表单的处理页用 url 模板标签表示 "{% url 'polls:vote' question.id %}"，表单以 post 的方式提交。
- forloop.counter 标签用于记录循环次数。
- 由于当前表单使用 post 方式提交数据，需要防止伪造的跨站点请求，表单中的 {% csrf_token %} 标签就可以解决这类问题。

接下来创建一个视图来接收并处理表单提交的信息：

```
#!/usr/bin/python
# -*- coding: UTF-8 -*-

from django.http import HttpResponse, HttpResponseRedirect
from django.shortcuts import get_object_or_404, render
from django.urls import reverse

from .models import Question, Choice

def vote(request, question_id):
    question = get_object_or_404(Question, pk=question_id)
    try:
        selected_choice = question.choice_set.get(pk=request.POST['choice'])
    except (KeyError, selected_choice.DoesNotExist):
        # Redisplay the question voting form.
        return render(request, 'polls/detail.html', {
            'question': question,
```

```
            'error_message': "还没有选择任何选项",
        })
    else:
        selected_choice.votes += 1
        selected_choice.save()
        # Always return an HttpResponseRedirect after successfully dealing
        # with POST data. This prevents data from being posted twice if a
        # user hits the Back button.
        return HttpResponseRedirect(reverse('polls:results', args=(question.id,)))
```

代码分析：

- request.POST 是一个类字典对象，以便通过关键字的名字获取提交的数据。如果表单提交的信息中不存在则抛出 KeyError。
- 信息处理结束后，使用 HttpResponseRedirect 方法跳转到新的页面以免用户单击浏览器后退按钮导致重新提交表单，这是一个很好的编程习惯。
- 为了防止在 HttpResponseRedirect 方法中使用 URL 硬编码，使用 reverse() 方法强制调用 URL 名字并且接收必要参数，而不是直接使用 URL。

由于提交问卷后页面会跳转到结果页，因此需要修改 results 视图：

```
def results(request, question_id):
    question = get_object_or_404(Question, pk=question_id)
    return render(request, 'polls/results.html', {'question': question})
```

新建 polls/results.html 模板并添加以下代码：

```
<h1>{{ question.question_text }}</h1>

<ul>
{% for choice in question.choice_set.all %}
    <li>{{ choice.choice_text }} -- 计票 {{ choice.votes }} {{ choice.votes| pluralize }} 次
{% endfor %}
</ul>

<a href="{% url 'polls:detail' question.id %}"> 重新投票? </a>
```

好了，到目前为止一个简单的投票系统就做好了，登录网站测试一下。

打开投票系统网址：http://127.0.0.1:8000/polls/，效果如下图所示。

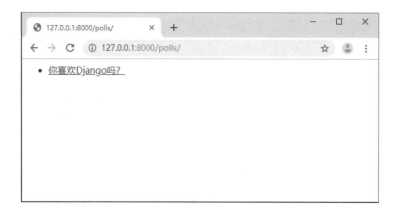

单击问卷链接，效果如下图所示。

选择任意选项，单击"提交"按钮，效果如下图所示。

3.13 添加样式

目前已经完成调查问卷系统的主要功能，但是还没有使用 CSS 样式对其进行美化，本节将带领读者学习如何在 Django 中使用 CSS 样式。

Django 将图片、脚本、样式表等文件称为静态文件（static files）。对小项目来说，如何处理静态文件不是什么值得注意的事情，你可以将它们放在任何地方，只要服务器能访问就可以了。但是，对大项目来说，尤其是包含很多应用程序的项目，处理每一个项目所使用的静态文件就比较困难了。

默认情况下，Django 会在应用程序根目录下查找 static 文件夹，这个文件夹就是用来存放静态文件的。

按照路径 polls/static/polls/style.css 创建一个 style.css 样式文件，具体 CSS 内容如下：

```
body {
    background-color: rgb(223, 204, 204);
}

li {
    color: green;
    text-decoration: none;
    list-style-type: decimal;
}
```

接下来修改模板 polls/templates/polls/index.html，在模板最顶部添加以下代码：

```
{% load static %}

<link rel="stylesheet" type="text/css" href="{% static 'polls/style.css' %}" />
```

{% static %} 标签用于生成静态文件的绝对路径。

到此为止，所有关于 CSS 的设置已经完成，重启 Web 服务，然后在浏览器中访问 index 页面，效果如下图所示。

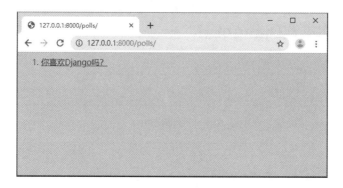

3.14 本地化

到此在线投票系统已经初步建立起来，但是页面显示上还有些缺陷，我们希望页面能够使用中文显示，因此还要对 settings.py 做以下设置：

❏ 配置时区。Django 的默认时区是" TIME_ZONE = 'UTC' "，将其修改为中国时区" TIME_ZONE = 'Asia/Shanghai' "。
❏ 配置语言。Django 的默认语言是英语" LANGUAGE_CODE = 'en-us' "，将其修改为简体中文" LANGUAGE_CODE = 'zh-hans' "。

> 提示
>
> Django 目前只支持部分语言的本地化，3.0 版本可接受的语言码请参考附录。

设置完成后，重启网站可以看到所有页面都已经显示中文。部分页面的显示效果如下面两幅图所示。

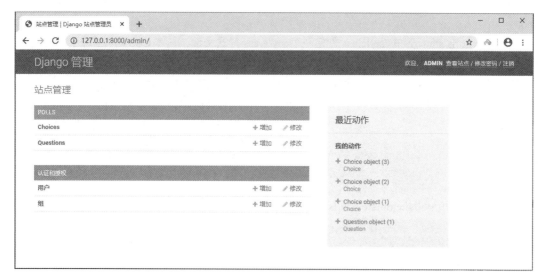

3.15 小结

到目前为止，我们已经学完所有的 Django 入门教程，包括安装和配置开发环境、创建项目和应用、配置数据库、开发视图和模板，并使用 CSS 样式对模板进行美化。

此时读者可以独立搭建 Web 应用了，但是，如果想要更愉快地使用 Django，还需要继续深入学习。前面提到的每一个知识点都可以扩展出很多内容，这部分内容是最基础的，强烈建议读者在学习阅读后面章节之前一定要完成本章内容的学习。

第 4 章
django-admin 和 manage.py

django-admin 是 Django 的命令行工具集，用于处理系统管理员相关操作，而 manage.py 是在创建 Django 项目的时候自动生成的，二者之间的作用完全一样。

django-admin 一般保存在环境变量中，在命令行或者终端都可以直接使用。它的物理路径在 Python 的 site-packages/django/bin 下，而 manage.py 存放在项目文件夹下。

django-admin 可以对不同的项目进行设置，但是需要在命令行中指定 --settings 参数或者修改 DJANGO_SETTINGS_MODULE 环境变量，而 manage.py 只对当前工程有效，可以直接拿来使用。

下面是工程 mysite 的 manage.py 脚本内容：

```python
#!/usr/bin/env python
"""Django's command-line utility for administrative tasks."""
import os
import sys


def main():
    os.environ.setdefault('DJANGO_SETTINGS_MODULE', 'mysite.settings')
    try:
        from django.core.management import execute_from_command_line
    except ImportError as exc:
        raise ImportError(
            "Couldn't import Django. Are you sure it's installed and "
            "available on your PYTHONPATH environment variable? Did you "
            "forget to activate a virtual environment?"
        ) from exc
    execute_from_command_line(sys.argv)


if __name__ == '__main__':
    main()
```

总之，manage.py 比 django-admin 更简单，本节所有命令都可以使用 manage.py 或者 python -m django 替代，以下是三个命令的等效用法：

```
$ django-admin <command> [options]
$ manage.py <command> [options]
$ python -m django <command> [options]
```

> **注意**
>
> 如果在运行某些 django-admin 命令的时候没有设置 --settings 参数，也没有设置环境变量 DJANGO_SETTINGS_MODULE，就可能会出现异常。例如，执行 check 命令检查项目基本情况时就会抛出以下异常：
>
> ```
> django.core.exceptions.ImproperlyConfigured: Requested setting INSTALLED_APPS,
> but settings are not configured. You must either define the environment variable
> DJANGO_SETTINGS_MODULE or call settings.configure() before accessing settings.
> ```
>
> 此时可以通过添加环境变量或者指定 --settings 和 --pythonpath 参数的方式来解决，可参考 check 命令的执行示例。

4.1 help

作用：取得帮助信息。

语法：

显示帮助信息以及可用命令。

```
django-admin help
```

显示可用命令列表。

```
django-admin help --commands
```

显示指定命令的详细帮助文档。

```
django-admin help <command>
```

示例：查看 check 命令的用法，如下图所示。

```
C:\WINDOWS\system32>django-admin help check
usage: django-admin check [-h] [--tag TAGS] [--list-tags] [--deploy]
                          [--fail-level {CRITICAL,ERROR,WARNING,INFO,DEBUG}]
                          [--version] [-v {0,1,2,3}] [--settings SETTINGS]
                          [--pythonpath PYTHONPATH] [--traceback] [--no-color]
                          [--force-color]
                          [app_label [app_label ...]]

Checks the entire Django project for potential problems.

positional arguments:
  app_label

optional arguments:
  -h, --help            show this help message and exit
  --tag TAGS, -t TAGS   Run only checks labeled with given tag.
  --list-tags           List available tags.
  --deploy              Check deployment settings.
  --fail-level {CRITICAL,ERROR,WARNING,INFO,DEBUG}
                        Message level that will cause the command to exit with
                        a non-zero status. Default is ERROR.
  --version             show program's version number and exit
  -v {0,1,2,3}, --verbosity {0,1,2,3}
                        Verbosity level; 0=minimal output, 1=normal output,
                        2=verbose output, 3=very verbose output
  --settings SETTINGS   The Python path to a settings module, e.g.
                        "myproject.settings.main". If this isn't provided, the
                        DJANGO_SETTINGS_MODULE environment variable will be
                        used.
  --pythonpath PYTHONPATH
                        A directory to add to the Python path, e.g.
                        "/home/djangoprojects/myproject".
  --traceback           Raise on CommandError exceptions
  --no-color            Don't colorize the command output.
  --force-color         Force colorization of the command output.
```

4.2 version

作用：取得当前 Django 版本信息。

语法：`django-admin version`

4.3 check

作用：检查工程中是否存在错误，默认会检查全部应用。

语法：`django-admin check [app_label [app_label ...]]`

示例：输入命令 `django-admin check auth polls --settings=mysite.settings --pythonpath=D:\Code\django3\mysite`

效果如下图所示。

```
D:\Code\django3\mysite>django-admin check auth polls --settings=mysite.settings --pythonpath=D:\Code\django3\mysite
System check identified no issues (0 silenced).
```

使用 --tag 参数可以约束 check 命令的检查范围，通过 -list-tags 参数查看全部可用的类别，如下图所示。

```
D:\Code\django3\mysite>django-admin check polls --settings=mysite.settings --pythonpath=D:\Code\django3\mysite --list-tags
admin
caches
database
models
staticfiles
templates
translation
urls

D:\Code\django3\mysite>django-admin check polls --settings=mysite.settings --pythonpath=D:\Code\django3\mysite --tag urls --tag models
System check identified no issues (0 silenced).
```

4.4 compilemessages

作用：将 .po 文件编译成用于国际化和本地化的 .mo 文件（使用 makemessages 命令可生成 .po 文件）。

语法：`django-admin compilemessages`

可选参数：

`--locale LOCALE, -l LOCALE`

指定待编译区域（locale），如果没有设置则编译全部区域的 .po 文件。

`--exclude EXCLUDE, -x EXCLUDE`

指定要从处理中排除的区域设置。如果没有提供，则不排除任何地区。

`--use-fuzzy, -f`

将模糊翻译编译到 .mo 文件。

`--ignore PATTERN, -i PATTERN`

编译消息文件时忽略与 glob 风格匹配的路径。可多次出现。

示例：首先在 polls 应用程序文件夹中准备下图所示的 .po 文件目录。

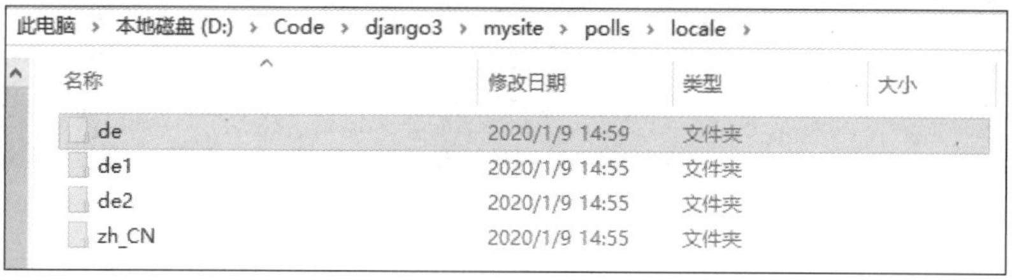

仅编译 de 区域的消息文件，如下图所示。

不编译 de 区域的消息文件，如下图所示。

因为 compilemessages 命令使用以下代码查找全部 locale 路径，所以 --ignore 参数所忽略的路径也应该包含在对应路径：

```
basedirs=[]
for dirpath, dirnames, filenames in os.walk('.', topdown=True):
    for dirname in dirnames:
        if dirname == 'locale':
            basedirs.append(os.path.join(dirpath, dirname))
basedirs
```

为了测试 --ignore 参数，在 mysite 文件夹下创建一个相同的 locale 文件夹，目录结构如下：

```
mysite/
    mysite/
            locale/
            ...
    polls/
            locale/
            ...
...
```

执行以上代码测试 compilemessages 命令所能查找的全部 locale 目录，如下图所示。

正常执行 compilemessages 命令，如下图所示。

```
D:\Code\django3\mysite>django-admin compilemessages --settings=mysite.settings --pythonpath=D:\Code\django3\mysite
processing file django.po in D:\Code\django3\mysite\polls\locale\de1\LC_MESSAGES
processing file django.po in D:\Code\django3\mysite\polls\locale\de2\LC_MESSAGES
processing file django.po in D:\Code\django3\mysite\polls\locale\de\LC_MESSAGES
processing file django.po in D:\Code\django3\mysite\polls\locale\zh_CN\LC_MESSAGES
processing file django.po in D:\Code\django3\mysite\mysite\locale\de1\LC_MESSAGES
processing file django.po in D:\Code\django3\mysite\mysite\locale\de2\LC_MESSAGES
processing file django.po in D:\Code\django3\mysite\mysite\locale\de\LC_MESSAGES
processing file django.po in D:\Code\django3\mysite\mysite\locale\zh_CN\LC_MESSAGES

D:\Code\django3\mysite>
```

可以看到根目录 mysite 下所有 locale 文件都被编译了，接下来使 --ignore 参数忽略 polls 目录下的 locale 文件，如下图所示。

```
D:\Code\django3\mysite>django-admin compilemessages --settings=mysite.settings --pythonpath=D:\Code\django3\mysite --ignore=polls\locale
processing file django.po in D:\Code\django3\mysite\mysite\locale\de2\LC_MESSAGES
processing file django.po in D:\Code\django3\mysite\mysite\locale\de1\LC_MESSAGES
processing file django.po in D:\Code\django3\mysite\mysite\locale\de\LC_MESSAGES
processing file django.po in D:\Code\django3\mysite\mysite\locale\zh_CN\LC_MESSAGES

D:\Code\django3\mysite>
```

可以看到，此时只编译了 mysite 目录下的 locale 文件。

4.5 createcachetable

作用：使用 settings 中的 CACHES 配置创建缓存表。Createcachetable 命令可接收两个参数。
- --database DATABASE：该参数用于指定创建缓存表的数据库，默认使用 default 数据库。
- --dry-run：打印出创建缓存表时所执行的 SQL 脚本，但是并不真正执行该脚本。

语法：`django-admin createcachetable`

示例：首先在 settings.py 中设置以下配置信息。

```
CACHES = {
    'default': {
        'BACKEND': 'django.core.cache.backends.db.DatabaseCache',
        'LOCATION': 'my_cache_table',
    }
}
```

执行 createcachetable 命令，如下图所示。

```
D:\Code\django3\mysite>django-admin createcachetable --settings=mysite.settings --pythonpath=D:\Code\django3\mysite --dry-run
CREATE TABLE "my_cache_table" (
    "cache_key" varchar(255) NOT NULL PRIMARY KEY,
    "value" text NOT NULL,
    "expires" datetime NOT NULL
);
CREATE INDEX "my_cache_table_expires" ON "my_cache_table" ("expires");

D:\Code\django3\mysite>
```

4.6 dbshell

作用：使用 settings 中配置的数据库连接串链接数据库并执行命令行。
语法：`django-admin dbshell`
示例：输入命令，如下图所示。

```
C:\WINDOWS\system32>django-admin dbshell --settings=mysite.settings --pythonpath=D:\Code\django2\mysite
SQLite version 3.7.14.1 2012-10-04 19:37:12
Enter ".help" for instructions
Enter SQL statements terminated with a ";"
sqlite> select * from polls_question;
1|你喜欢Django吗?|2019-09-17 00:43:51
sqlite>
```

4.7　diffsettings

作用：显示当前工程的配置文件与 Django 的默认配置文件之间的差别。如果默认配置文件中缺少某项配置则显示"###"。默认输出格式（--hash）比较乱，可以使用 --output unified 进行格式化，此时默认配置前会以减号(-)开头。

语法：`django-admin diffsettings`

示例：输入命令，如下图所示。

```
C:\WINDOWS\system32>django-admin diffsettings --output unified --settings=mysite.settings --pythonpath=D:\Code\django2\mysite
- AUTH_PASSWORD_VALIDATORS = []
+ AUTH_PASSWORD_VALIDATORS = [{'NAME': 'django.contrib.auth.password_validation.UserAttributeSimilarityValidator'}, {'NAME': 'django.contrib.auth.password_validation.MinimumLengthValidator'}, {'NAME': 'django.contrib.auth.password_validation.CommonPasswordValidator'}, {'NAME': 'django.contrib.auth.password_validation.NumericPasswordValidator'}]
+ BASE_DIR = 'D:\\Code\\django2\\mysite'
- DATABASES = {}
+ DATABASES = {'default': {'ENGINE': 'django.db.backends.sqlite3', 'NAME': 'D:\\Code\\django2\\mysite\\db.sqlite3', 'ATOMIC_REQUESTS': False, 'AUTOCOMMIT': True, 'CONN_MAX_AGE': 0, 'OPTIONS': {}, 'TIME_ZONE': None, 'USER': '', 'PASSWORD': '', 'HOST': '', 'PORT': '', 'TEST': {'CHARSET': None, 'COLLATION': None, 'NAME': None, 'MIRROR': None}}}
- DEBUG = False
+ DEBUG = True
```

4.8　dumpdata

作用：在标准输出设备中输出指定应用程序数据，输出的数据可用于 loaddata 命令。

语法：`django-admin dumpdata [app_label[.ModelName] [app_label[.ModelName] ...]]`

示例 1：使用 --indent 参数格式化 json 字符串。输入命令，如下图所示。

```
C:\WINDOWS\system32>django-admin dumpdata polls.question --indent=4 --settings=mysite.settings --pythonpath=D:\Code\django2\mysite
[
{
    "model": "polls.question",
    "pk": 1,
    "fields": {
        "question_text": "\u4f60\u559c\u6b22Django\u5417\uff1f",
        "pub_date": "2019-09-17T00:43:51Z"
    }
}
]
```

示例 2：使用 --output 将 polls_question 表中的数据输出到文件。输入命令，如下图所示。

```
C:\WINDOWS\system32>django-admin dumpdata polls.question --output=D:\Code\django2\mysite\question.json --indent=4 --settings=mysite.settings --pythonpath=D:\Code\django2\mysite
[.........................................................]
C:\WINDOWS\system32>
```

4.9 flush

作用:清空当前数据库中的数据,但是会保留 migration 的变更。如果确实想要一个干净的数据库,Django 官方推荐通过删除数据库并重建的方式而不是清空数据库。

语法:`django-admin flush`

示例:输入命令,如下图所示。

```
D:\Code\django3\mysite>django-admin flush --settings=mysite.settings --pythonpath=D:\Code\django3\mysite
You have requested a flush of the database.
This will IRREVERSIBLY DESTROY all data currently in the 'D:\\Code\\django3\\mysite\\db.sqlite3' database,
and return each table to an empty state.
Are you sure you want to do this?

    Type 'yes' to continue, or 'no' to cancel: no
Flush cancelled.

D:\Code\django3\mysite>
```

如果在交互窗口输入 yes,那么命令执行结束,将会清空现有数据库仅保留最新的数据库结构。

4.10 inspectdb

作用:查询数据库表或者视图对应的 Django 模型。如果没有提供参数,则仅在使用 --include-views 选项时为视图创建模型。

通过该命令可以很方便地将已有的数据库表或者视图转换成 Django 模型。

语法:`django-admin inspectdb [table [table ...]]`

示例 1:查看 polls_question 表对应的 Django 模型,输入命令,如下图所示。

```
D:\Code\django3\mysite>django-admin inspectdb polls_question --settings=mysite.settings --pythonpath=D:\code\django3\mysite
# This is an auto-generated Django model module.
# You'll have to do the following manually to clean this up:
#   * Rearrange models' order
#   * Make sure each model has one field with primary_key=True
#   * Make sure each ForeignKey and OneToOneField has `on_delete` set to the desired behavior
#   * Remove `managed = False` lines if you wish to allow Django to create, modify, and delete the table
# Feel free to rename the models, but don't rename db_table values or field names.
from django.db import models


class PollsQuestion(models.Model):
    question_text = models.CharField(max_length=200)
    pub_date = models.DateTimeField()

    class Meta:
        managed = False
        db_table = 'polls_question'

D:\Code\django3\mysite>
```

示例 2:查看数据库视图对应的 Django 模型。

首先在数据库中创建一个视图,该视图能显示调查问卷及问卷答案,视图对应的 SQL 脚本如下:

```
select q.question_text, q.pub_date, c.choice_text, c.votes
from polls_question as q inner join polls_choice as c on q.id = c.question_id
```

新视图在数据库中的显示如下图所示。

第 4 章　django-admin 和 manage.py

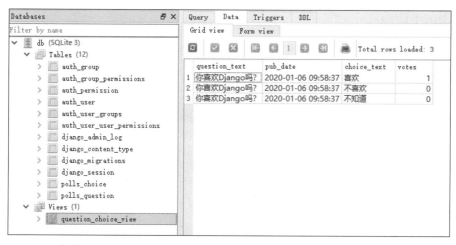

执行 inspectdb 命令，如下图所示。

```
D:\Code\django3\mysite>django-admin inspectdb question_choice_view --settings=mysite.settings --pythonpath=D:\code\django3\mysite --include-views
# This is an auto-generated Django model module.
# You'll have to do the following manually to clean this up:
#   * Rearrange models' order
#   * Make sure each model has one field with primary_key=True
#   * Make sure each ForeignKey and OneToOneField has `on_delete` set to the desired behavior
#   * Remove `managed = False` lines if you wish to allow Django to create, modify, and delete the table
# Feel free to rename the models, but don't rename db_table values or field names.
from django.db import models


class QuestionChoiceView(models.Model):
    question_text = models.CharField(max_length=200, blank=True, null=True)
    pub_date = models.DateTimeField(blank=True, null=True)
    choice_text = models.CharField(max_length=200, blank=True, null=True)
    votes = models.IntegerField(blank=True, null=True)

    class Meta:
        managed = False  # Created from a view. Don't remove.
        db_table = 'question_choice_view'

D:\Code\django3\mysite>
```

注意

- 当 inspectdb 命令无法将数据库中的字段类型转换为 Django 模型字段类型时将会使用 TextField，同时插入一条注释：'This field type is a guess.'。
- 如果数据库字段名是 Python 保留字，如 class、pass，那么 inspectdb 命令会为模型属性名自动添加一个"_field"后缀。
- inspectdb 命令不会根据数据库字段默认值生成 model 字段的默认值。
- 默认情况下 inspectdb 命令生成的模型是非 Django 托管模型（managed=False），如果想生成托管模型，可以使用 managed 参数。

针对不同数据库需要注意以下几点：

- 在 Oracle 中使用 --include-views 可以为物化视图（materialized view）生成模型。
- 在 PostgreSQL 中，inspectdb 命令可以为外部表（foreign table）生成模型，使用 --include-views 可以为物化视图（materialized view）生成模型，使用 --include-partitions 可以为部分表（partition table）生成模型。

4.11 loaddata

作用:加载数据到数据库。

语法:`django-admin loaddata fixture [fixture ...]`

示例:创建 fixture 文件 question.json,将该文件放在 manage.py 同级目录,文件内容如下。

```
[{
    "model": "polls.question",
    "pk": 2,
    "fields": {
        "question_text": "Django是一套强大的Web开发框架",
        "pub_date": "2020-01-20T15:31:00Z"
    }
}]
```

执行 loaddata 命令,如下图所示。

```
D:\Code\django3\mysite>django-admin loaddata question.json --settings=mysite.settings --pythonpath=D:\code\django3\mysite
Installed 1 object(s) from 1 fixture(s)

D:\Code\django3\mysite>
```

执行结果如下图所示。

id	question_text	pub_date
1	你喜欢Django吗?	2020-01-06 09:58:37
2	Django是一套强大的Web开发框架	2020-01-20 15:31:00

> **注意**
>
> fixture 是序列化好的数据文件,文件格式包含 json 和 xml。Django 只能从以下 3 类位置查找 fixture:
> - 应用程序下的 fixtures 文件夹。
> - 配置文件中 FIXTURE_DIRS 指定的路径。
> - fixture 文件路径。
>
> Loaddata 命令可以在压缩文件中查找 fixture,Loaddata 支持 zip、gz 和 bz2 等压缩格式,例如执行以下命令:
>
> `django-admin loaddata mydata.json`
>
> 该命令将会同时查找 mydata.json、mydata.json.zip、mydata.json.gz 或者 mydata.json.bz2,直到找到第一个 mydata.json 文件。
>
> 需要注意的是,不能将相同名字不同类型的 fixture 文件放在同一路径下,如 mydata.json 和 mydata.xml。

4.12 makemessages

作用:查找整个源代码路径以找出全部翻译字符串并生成一个新的消息文件或者更新已有的消息文件。

语法:`django-admin makemessages`

在介绍代码示例前,首先修改 index.html,增加翻译字符串:

```
{% load i18n %}
{% load static %}
<link rel="stylesheet" type="text/css" href="{% static 'polls/style.css' %}">

{% if latest_question_list %}
    <ul>
    {% for question in latest_question_list %}
        <li><a href="{% url 'polls:detail' question.id %}" >{{ question.question_text }}</a></li>
    {% endfor %}
    </ul>
{% else %}
    <p>{% trans "warningmsg" %}</p>
{% endif %}
```

然后在 polls\templates\polls\ 文件夹下新建一个脚本文件 trans.js,文件内容如下(该脚本没有实际意义,仅用于演示 makemessages 命令):

```
(function() {
    window.alert(gettext("warningmsg"));
})();
```

示例 1:生成中文翻译文件,如下图所示。

```
D:\Code\django3\mysite>django-admin makemessages -l zh_CN --settings=mysite.settings --pythonpath=D:\code\django3\mysite
processing locale zh_CN

D:\Code\django3\mysite>
```

示例 2:使用 domain 参数为 js 脚本生成中文翻译文件。

Domain 参数可接受两个值:

- django:该参数用于从 *.py、*.html 和 *.txt 文件中查找翻译字符串。
- djangojs:该参数用于从 *.js 文件中查找翻译字符串,当指定 djangojs 参数时,应保证 js 文件位于 tempaltes 文件夹下,否则需要在模板中使用外链引用该脚本文件。

```
D:\Code\django3\mysite>django-admin makemessages -l zh_CN -d djangojs --settings=mysite.settings --pythonpath=D:\code\django3\mysite
processing locale zh_CN

D:\Code\django3\mysite>
```

4.13 startproject

作用:创建 Django 项目。

语法:`django-admin startproject name [directory]`

命令默认在当前目录创建一个文件夹,文件夹下包含 manage.py 文件以及工程文件夹,在工程文件夹下包含 settings.py 文件和其他必要文件。

4.14 startapp

作用:创建 Django 应用程序。

语法：django-admin startapp name [directory]

可选参数：

--template TEMPLATE

导入外部应用程序模板，TEMPLATE 可以是包含模板文件的路径、包含压缩包的路径或者 URL。例如，下面命令会将 my_app_template 路径下的模板文件复制到 myapp 应用程序中：

django-admin startapp --template=/Users/jezdez/Code/my_app_template myapp

而下面命令会将 GitHub 上其他项目的模板复制到 myapp 应用中：

django-admin startapp --template=https://github.com/ebertti/django-registration-bootstrap/archive/master.zip myapp

4.15 runserver

作用：在当前机器上启动一个轻量级的 Web 服务器，默认服务器端口号是 8000。

语法：django-admin runserver [addrport]

示例：

```
django-admin runserver
django-admin runserver 1.2.3.4:8000
django-admin runserver 7000
django-admin runserver [2001:0db8:1234:5678::9]:7000
```

4.16 sendtestemail

作用：发送测试邮件以检测邮箱设置是否正确。

语法：django-admin sendtestemail foo@example.com bar@example.com

示例：本例以 QQ 邮件服务为例讲解如何在 Django 中发送邮件。首先在 settings.py 中设置 QQ 邮件服务：

```
EMAIL_HOST = 'smtp.qq.com'
EMAIL_HOST_USER = '***@qq.com'
EMAIL_HOST_PASSWORD = 'QQ邮件服务授权码'
EMAIL_USE_TLS = True

DEFAULT_FROM_EMAIL = '李健 <***@qq.com>'
ADMINS = (
    ('李健', '***@qq.com'),
)

MANAGERS = ADMINS
```

> **注意**
>
> QQ 邮件服务授权码可以在 QQ 邮箱的设置中获取。

完成以上配置后执行以下命令测试邮件功能：

```
python manage.py sendtestemail 目标邮箱地址
```

下图是收到的测试邮件内容。

4.17 shell

作用：启动一个 Python 交互窗口。
语法：

```
django-admin shell --interface {ipython,bpython,python}
django-admin shell --i {ipython,bpython,python}
```

> **注意**
>
> 默认情况下，Django 使用 ipython 或者 bpython 启动交互模式，如果同时安装了这两个交互工具，那么在执行 shell 命令的时候需要指定具体使用哪一个。
> 可以使用 pip 安装以上交互工具，例如安装 ipython：pip install ipython。

示例：输入命令，如下图所示。

```
D:\Code\django3\mysite>python manage.py shell
Python 3.8.0 (tags/v3.8.0:fa919fd, Oct 14 2019, 19:21:23) [MSC v.1916 32 bit (Intel)] on win32
Type "help", "copyright", "credits" or "license" for more information.
(InteractiveConsole)
>>>
```

4.18 迁移

Django 通过迁移命令将 Model 中的任何修改写入到数据库中，例如增加新模型、修改已有模型字段等。

4.18.1 makemigrations

作用：根据模型的变化生成对应的迁移代码，该代码用于更新数据库。
语法：django-admin makemigrations [app_label [app_label ...]]
如果没有填写任何参数，Django 会检查所有应用程序中的模型并生成迁移脚本，脚本存放在每个应用下面一个叫作 migrations 的文件夹下，脚本名字类似 0001_initial.py 格式。
示例 1：为一个应用程序生成迁移。输入命令，如下图所示。

```
D:\Django\demo\mysite>python manage.py makemigrations blog
Migrations for 'blog':
  blog\migrations\0003_test.py
    - Create model Test

D:\Django\demo\mysite>
```

示例 2：生成一个空迁移，高级开发人员可以在空迁移中编写代码。输入命令，如下图所示。

```
D:\Code\django3\mysite>python manage.py makemigrations polls --empty
Migrations for 'polls':
  polls\migrations\0003_auto_20200121_1456.py

D:\Code\django3\mysite>
```

示例 3：生成一个带名称的迁移。输入命令，如下图所示。

```
D:\Code\django3\mysite>python manage.py makemigrations polls --empty --name empty
Migrations for 'polls':
  polls\migrations\0004_empty.py

D:\Code\django3\mysite>
```

4.18.2 migrate

作用：将模型的最新状态部署到数据库。

语法：`django-admin migrate [app_label] [migration_name]`

如果执行 migrate 命令时没有给出任何参数，Django 会将系统中所有应用程序模型的更改部署到数据库。

如果执行 migrate 命令时指定了应用程序名，Django 仅将指定的应用程序的模型修改部署到数据库。注意，如果该应用程序的模型与其他应用程序模型之间存在关联，那么其他关联的应用程序模型的修改也可能被部署到数据库。

如果执行 migrate 命令的同时给出了应用程序名和 migration 名字，系统将会把数据库恢复到一个该迁移前的版本，该操作常用于回滚错误的数据库变更。

为测试回滚操作，在 polls\models.py 中新建一个模型，并使用 migrate 命令将其应用到数据库中。模型代码如下：

```
class Rolback(models.Model):
    name = models.CharField(max_length=200)
```

所有 migration 信息保存在 django_migrations 数据表中，下图是一段时间内发生的数据迁移操作。

16	16	auth	0011_update_proxy_permissions	2020-01-06 08:17:51.940098
17	17	sessions	0001_initial	2020-01-06 08:17:52.015224
18	18	polls	0001_initial	2020-01-06 09:40:28.980627
19	19	polls	0002_auto_20200121_1429	2020-01-21 07:20:14.503226
20	20	polls	0003_auto_20200121_1456	2020-01-21 07:20:14.580957
21	21	polls	0004_empty	2020-01-21 07:20:14.646783
22	22	polls	0005_rolback	2020-01-21 07:20:14.727345

其中，最后一行（0005_rolback）就是前面用于生成测试模型的迁移记录，此时执行下面代码将会撤销本次迁移（0004_empty 是 0005_rolback 的前一次操作）：

```
python manage.py migrate polls 0004_empty
```

输入命令，如下图所示。

刷新数据库，可以看到 0005_rolback 已经被删除了。

如果想撤销所有数据库更改，可以使用 zero 代替 migrationname。

其他可选参数如下。

- --fake：对于高级用户，仅仅想设置当前的 migration 状态，并不需要真正去更新数据库，例如已经手工更新过数据库，此时可以使用 fake 参数。

```
python manage.py migrate polls --fake
```

- --database DATABASE：将模型更改应用到指定的数据库，默认情况会更新到 settings.py 里面的 default 数据库。

4.18.3 sqlmigrate

作用：输出某一个 migrate 对应的 SQL 语句。

语法：django-admin sqlmigrate app_label migration_name

示例：打印出初始化的 SQL 脚本。输入命令，如下图所示。

4.18.4 showmigrations

作用：显示 migrations 记录。

语法：django-admin showmigrations [app_label [app_label ...]]

可以通过 --list 或者 --plan 参数设置显示格式。--list 按照应用程序显示 migration 记录，该参数缩写为 -l，--plan 显示所有记录，缩写为 -p。如果某一次 migration 已经被部署到数据库中，在该记录前就会显示 [X]，否则显示 [ ]。

示例：输入命令，如下图所示。

```
D:\Django\demo\mysite>python manage.py showmigrations
admin
 [X] 0001_initial
 [X] 0002_logentry_remove_auto_add
auth
 [X] 0001_initial
 [X] 0002_alter_permission_name_max_length
 [X] 0003_alter_user_email_max_length
 [X] 0004_alter_user_username_opts
 [X] 0005_alter_user_last_login_null
 [X] 0006_require_contenttypes_0002
 [X] 0007_alter_validators_add_error_messages
 [X] 0008_alter_user_username_max_length
 [X] 0009_alter_user_last_name_max_length
blog
 [X] 0001_initial
 [X] 0002_auto_20171221_1726
 [ ] 0003_test
contenttypes
 [X] 0001_initial
 [X] 0002_remove_content_type_name
polls
 [X] 0001_initial
sessions
 [X] 0001_initial

D:\Django\demo\mysite>
```

4.19 changepassword

作用：修改用户密码。

语法：django-admin changepassword [<username>]

示例：输入命令，如下图所示。

```
C:\WINDOWS\system32>django-admin changepassword admin --settings=mysite.settings --pythonpath=D:\Code\django 2\mysite
Changing password for user 'admin'
Password:
Password (again):
Password changed successfully for user 'admin'
```

> **注意**
>
> 只有安装了 Django 的认证系统（django.contrib.auth），changepassword 命令才可用。

4.20 createsuperuser

作用：为 Django 系统创建一个超级用户。

语法：django-admin createsuperuser

4.21 collectstatic

作用：收集静态文件并保存到 STATIC_ROOT 所指定的文件夹。

语法：django-admin collectstatic

示例：输入命令，如下图所示。

4.22 findstatic

作用：查找一个或多个静态文件路径。
语法：`django-admin findstatic staticfile [staticfile ...]`
示例：输入命令，如下图所示。

4.23 默认选项

学习前面的命令之后可以看到某些参数是在所有命令中都可用的，如 --settings，这样的参数称作默认选项（default options），下面具体介绍一下类似的参数。

- --pythonpath PYTHONPATH：将指定路径添加到 Python path，如果没有指定这个参数，django-admin 将会使用 PYTHONPATH 环境变量。
- --settings SETTINGS：指定 Django 应用程序的 settings 模块，参数值以 Python 路径的语法格式书写，例如 mysite.settings。如果没有给出 settings 选项，那么 django-admin 会查找 DJANGO_SETTINGS_MODULE 环境变量。
- --traceback：当出现 CommandError 异常时显示一个完整的堆栈跟踪。
- --verbosity：指定输出到控制台的提示信息或者调试信息的数量，可选等级如下。
  - 0：不输出任何信息。
  - 1：正常输出信息（默认值）。
  - 2：输出详细信息。
  - 3：输出非常详细的信息。

示例：django-admin migrate --verbosity 2
- --no-color：禁用彩色显示。
- --force-color：Django 2.2 新增命令。强制为命令行输出信息着色，但是，当命令行窗口不支持时仍采用默认方式显示。
- --skip-checks：Django 3.0 新增命令。在运行命令之前跳过系统检查。只有在 requires_system_checkcommand 属性设置为 True 时，此选项才可用。

第 5 章

配置

在搭建 Django 应用程序的时候需要进行一定的配置，除了前面使用过的数据库配置、系统语言配置外，Django 还提供了更多的配置项，这些配置项都存放在配置文件中。

Django 的配置文件是一个 Python 模块，所有配置项都是模块级别的变量。本章将详细介绍 Django 配置。

5.1 Django 配置文件

由于 Django 的配置文件是一个 Python 模块，因此必须遵循以下原则。

❑ 不能够出现 Python 语法错误。

❑ 可以使用 Python 语法动态指定配置值，如：

```
MY_SETTING = [str(i) for i in range(30)]
```

❑ 可以从其他配置文件中引入变量。

使用 Django 时必须通过环境变量 DJANGO_SETTINGS_MODULE 指定当前工程所使用的配置文件，默认情况下在 manage.py 中指定配置文件，如：

```python
#!/usr/bin/env python
"""Django's command-line utility for administrative tasks."""
import os
import sys


def main():
    os.environ.setdefault('DJANGO_SETTINGS_MODULE', 'mysite.settings')
    try:
        from django.core.management import execute_from_command_line
    except ImportError as exc:
        raise ImportError(
            "Couldn't import Django. Are you sure it's installed and "
            "available on your PYTHONPATH environment variable? Did you "
            "forget to activate a virtual environment?"
        ) from exc
    execute_from_command_line(sys.argv)


if __name__ == '__main__':
    main()
```

如果使用 WSGI 部署 Django 应用程序，需要在 wsgi.py 中指定配置文件，如：

```
import os
from django.core.wsgi import get_wsgi_application
os.environ.setdefault("DJANGO_SETTINGS_MODULE", "mysite.settings")
application = get_wsgi_application()
```

如果配置文件缺少对某个配置的设置，则会使用默认值，每一个配置项在 Django 的默认配置文件中都已经给出了默认值。Django 的默认配置文件路径是 django/conf/global_settings.py。

Django 按照下面算法编译配置文件：

- 加载 global_settings.py。
- 加载工程指定的配置文件，使用工程指定的配置重写默认值。

5.1.1　引用 Django 配置信息

在 Django 应用程序中可以通过导入 django.conf.settings 来引用 Django 配置信息，例如：

```
from django.conf import settings
if settings.DEBUG:
    pass
```

需要注意的是，django.conf.settings 是一个 Python 对象，不是模块，不能使用下面方法单独导入配置信息：

```
from django.conf.settings import DEBUG
```

另外，由于配置文件是在 Django 编译时加载的，因此不能在运行时对系统配置进行修改，也不能指定是导入 global_settings 还是其他配置文件。

还有一种特殊的引用配置信息的方式，那就是忽略 DJANGO_SETTINGS_MODULE。某些情况下可能只是使用 Django 项目的一部分代码，那么此时不需要引用整个配置文件，可以使用以下代码手工配置 Django，这种情况很少用，只要了解就可以了：

```
from django.conf import settings
settings.configure(DEBUG=True)
```

configure 方法可以接收任意关键字参数，每个参数都代表一个 Django 配置，同时 configure 方法还可以重新指定默认配置文件：

```
from django.conf import settings
from myapp import myapp_defaults

settings.configure(default_settings=myapp_defaults, DEBUG=True)
```

需要注意的是，DJANGO_SETTINGS_MODULE 和 configure() 只能使用一个，不能同时使用也不能同时不使用。

5.1.2　django.setup

如果只使用 Django 的一部分代码而不启动整个项目，例如加载某个模板并渲染它，或者使用 ORM 来加载一些数据，此时可以使用 django.setup() 来加载配置文件，例如：

```
import django
from django.conf import settings
```

```
import os
import sys

os.environ.setdefault('DJANGO_SETTINGS_MODULE', 'mysite.settings')
django.setup()

from polls.urls import *
print(urlpatterns)
```

使用 django.setup() 需要注意以下几点:
- 被引用代码必须是独立存在的（standalone）。
- django.setup() 只会被调用一次，在使用 django.setup() 的时候不要引用复杂的代码，例如可重用的业务代码。

5.2 Cache

5.2.1 CACHES

默认值:

```
{
    'default': {
        'BACKEND': 'django.core.cache.backends.locmem.LocMemCache',
    }
}
```

这是一个嵌套的字典对象，定义了项目中可能用到的全部缓存，default 是缓存别名，BACKEND 是缓存的解释引擎。

CACHES 至少要定义一个 default 缓存，可以另外指定任意多个可选缓存。

可选的 BACKEND 包括以下几种:
- 'django.core.cache.backends.db.DatabaseCache'
- 'django.core.cache.backends.dummy.DummyCache'
- 'django.core.cache.backends.filebased.FileBasedCache'
- 'django.core.cache.backends.locmem.LocMemCache'
- 'django.core.cache.backends.memcached.MemcachedCache'
- 'django.core.cache.backends.memcached.PyLibMCCache'

其他可选参数介绍如下。

1. KEY_PREFIX:
默认值：''（空字符串）
这个字符串会被自动添加在 Djagno 生成的缓存的 Key 前面作为前缀。

2. LOCATION:
默认值：''（空字符串）
表示缓存地址。如果是文件缓存，那么 location 是文件目录；如果是 memcache，那么 location 是 memcache 服务器的主机名和端口号，或者是本地内存缓存的名字。

下面是文件缓存的例子：

```
CACHES = {
    'default': {
        'BACKEND': 'django.core.cache.backends.filebased.FileBasedCache',
        'LOCATION': '/var/tmp/django_cache',
    }
}
```

3. OPTIONS:

默认值：None

传递给 BACKEND 的额外参数，不同的 BACKEND 所支持的参数也是不同的。下面是实现自有淘汰策略的缓存后端所使用的两个选项：

- MAX_ENTRIES：删除旧值之前允许缓存的最大条目。默认是 300。
- CULL_FREQUENCY：当达到 MAX_ENTRIES 时被淘汰的部分条目。实际比率为 1/CULL_FREQUENCY；当达到 MAX_ENTRIES 时，设置为 2 会淘汰一半的条目。这个参数应该是一个整数，默认为 3。

CULL_FREQUENCY 的值为 0 意味着当缓存数达到 MAX_ENTRIES 时，整个缓存都会被清空。在一些缓存系统中（尤其是数据库），这会通过降低缓存命中率为代价来更快地淘汰旧缓存。

4. TIMEOUT:

默认值：300

缓存过期时间，单位为秒。如果 TIMEOUT 设置为 None，那么缓存永不过期。

5. VERSION:

默认值：1

通过 Django 生成的缓存的默认版本号。

5.2.2 CACHE_MIDDLEWARE_ALIAS:

默认值：default

缓存中间件所使用的缓存。

5.2.3 CACHE_MIDDLEWARE_KEY_PREFIX:

默认值：''（空字符串）

缓存中间件所生成的缓存 Key 的前缀。这个前缀包含 KEY_PREFIX 的设置内容，但是并不替换 KEY_PREFIX。

5.2.4 CACHE_MIDDLEWARE_SECONDS:

默认值：600

缓存中间件缓存一个页面的默认时长，单位为秒。

5.3 数据库

5.3.1 DATABASES

默认值：{}

DATABASES 用于指定网站所使用的数据库类型以及连接方式，它是一个嵌套的字典对象，最外层的 Key 是数据库别名，值是具体数据库的配置信息。

DATABASES 必须包含一个别名为 default 的数据库配置，同时可以额外设置任意多个数据库。Django 支持包括 PostgreSQL、MySQL、SQLite、Oracle 等几种主流数据库。默认为一个空字典，此时会使用 SQLite 数据库，该数据库在创建 Django 应用程序时被自动创建。这就是在第一次搭建 Django 网站的时候，什么都没有做，就可以添加管理员并且使用后台管理系统的原因，此时所有信息都存在这个 SQLite 数据库中了。

默认配置等价于：

```
DATABASES = {
    'default': {
        'ENGINE': 'django.db.backends.SQLite3',
        'NAME': os.path.join(BASE_DIR, 'db.SQLite3'),
    }
}
```

如果需要链接到其他数据库，那么需要给出更多的配置信息。下面是一个用于链接 MySQL 数据库的配置信息：

```
DATABASES = {
    'default': {
        'ENGINE': 'django.db.backends.mysql',
        'NAME': 'polls',
        'USER': 'root',
        'PASSWORD': '',
        'HOST': '',
        'PORT': '',
    }
}
```

从上面的配置信息可以知道，Django 与其他语言链接数据库的方式相似，同样需要链接数据库的用户名、密码，同时需要给出数据库所在的主机名、端口号以及数据库名。

下面的参数可能在更复杂的配置信息中被用到。

1. ATOMIC_REQUESTS:

默认值：False

如果设置为 True，将会把每个视图的数据库操作封装到一个数据库事务中。

2. AUTOCOMMIT:

默认值：True

如果设置为 False，那么将会终止 Django 默认的事务管理功能。

3. ENGINE:

默认值：''（空字符串）

数据库引擎。针对不同的数据库，Django 提供了不同的引擎：
- 'django.db.backends.postgresql'
- 'django.db.backends.mysql'
- 'django.db.backends.SQLite3'
- 'django.db.backends.oracle'

如果以上数据库都不是你想要的，那么可以引用自己的数据库引擎。使用时只要将 ENGINE 指向自定义数据库引擎文件就可以了，例如 mypackage.backends.whatever。

4. HOST:

默认值：''（空字符串）

数据库所在主机名，如果值为空表示本机。SQLite 不需要指定该参数。

如果使用 TCP 的方式链接数据库，可以使用"localhost"或者"127.0.0.1"。

5. NAME:

默认值：''（空字符串）

数据库名，对于 SQLite 数据库，需要给定 SQLite 文件路径，不论是 Windows 系统还是 Linux 系统，这个文件路径中一律使用斜杠"/"，例如 C:/homes/user/mysite/SQLite3.db。

6. CONN_MAX_AGE:

默认值：0

数据库会话的生命周期，单位为秒。默认值为 0，表示每次请求结束立刻关闭数据库链接。如果设置为 None，那么就没有限制，这样不安全，也容易占用有限的数据库资源，不建议使用这种方式。

7. OPTIONS:

默认值：{}

连接数据库时需要使用的附加参数，不同的数据库引擎所支持的参数不相同。

8. USER:

默认值：''（空字符串）

连接数据库的用户名。SQLite 不需要指定。

9. PASSWORD:

默认值：''（空字符串）

连接数据库的用户密码。SQLite 不需要指定。

10. PORT:

默认值：''（空字符串）

为数据库开放的端口号，如果值为空表示默认端口。SQLite 不需要指定。

11. TIME_ZONE:

默认值：None

数据库中所使用的时区。DATABASES 中的 TIME_ZONE 参数与配置文件中的 TIME_ZONE 参数可以使用的值完全一样，只是 DATABASES 中的 TIME_ZONE 只是用来表示数据库中的数据存储格式。

这样做可以很方便地使用那些非 UTC 时间存储的数据库。为了避免由于变更时区出现问题，不应该改变 Django 托管的数据库时区。

如果 USE_TZ 参数设置为 True 并且正在使用不支持时区设置的数据库（如 SQLite、MySQL、Oracle），Django 会按照 DATABASES 中的 TIME_ZONE 来读写数据，如果 DATABASES 中没有设置 TIME_ZONE，那么 Django 会使用 UTC 时区来读写数据。

如果 USE_TZ 参数设置为 True 并且正在使用支持时区的数据库（如 PostgreSQL），那么设置这个参数是错误的。

如果 USE_TZ 参数设置为 False 也不应该设置这个参数。

12. DISABLE_SERVER_SIDE_CURSORS：

默认值：`False`

如果希望在 QuerySet.iterator() 中禁用服务器端游标，可以将这个参数设置为 True。

这个参数只对 PostgreSQL 数据库有效。

13. TEST：

默认值：`{}`

用于配置测试数据库。

下面是一个配置了测试数据库的例子：

```
DATABASES = {
    'default': {
        'ENGINE': 'django.db.backends.postgresql',
        'USER': 'mydatabaseuser',
        'NAME': 'mydatabase',
        'TEST': {
            'NAME': 'mytestdatabase',
        },
    },
}
```

下面是 TEST 字典所支持的关键字。

[CHARSET]：

默认值：`None`

创建数据库时所使用的字符编码。不同数据库所支持的字符编码不同，如 PostgreSQL 支持 BIG5、EUC_CN、GBK，MySQL 支持 big5、gbk、gb2312 等。

[COLLATION]：

默认值：`None`

创建数据库时所使用的排序规则，目前只有 MySQL 支持。下面是中文常用的排序规则：

Big5: big5_chinese_ci

gb2312: gb2312_chinese_ci

gbk: gbk_chinese_ci

[DEPENDENCIES]：

默认值：`['default']`

创建数据库时的依赖顺序。

> **注意**
> default 数据库是没有依赖的,不能设置 DEPENDENCIES 参数。

[MIRROR]:

默认值: None

测试过程中需要进行镜像的数据库别名。该参数的存在使得测试主从复制数据库变得可能了。

[NAME]:

默认值: None

执行测试套件时所使用的数据库名字。

默认情况下,SQLite 使用一个内存数据库,其他数据库使用的数据库名字按 'test_' + DATABASE_NAME 的形式生成。

[SERIALIZE]:

布尔值,用于规定默认测试器是否在执行测试前将数据库序列化为一个 JSON 内存对象。如果测试用例没有设置 serialized_rollback=True,那么可以将该参数设置为 False 以加快测试创建时间。

[TEMPLATE]:

PostgreSQL 数据库特有参数,用于指定创建测试数据库时所使用的模板。

[CREATE_DB]:

默认值: True

Oracle 数据库特有参数,如果设置为 False,那么在测试开始前不会自动创建表空间,也不会在测试束后删除表空间。

[CREATE_USER]:

默认值: True

Oracle 数据库特有参数,如果设置为 False,那么在测试开始前不会自动创建测试用户,也不会在测试结束后删除测试用户。

[USER]:

默认值: None

Oracle 数据库特有参数,执行测试时所使用的数据库用户名,如果没有指定,Django 会使用 'test_' + USER。

[PASSWORD]:

默认值: None

Oracle 数据库特有参数,执行测试时所使用的数据库用户密码,如果没有指定,Django 会使用一个随机密码。

[ORACLE_MANAGED_FILES]:

默认值: False

Oracle 数据库特有参数,如果设置为 True 将会使用 OMF(Oracle Managed Files)忽略 DATAFILE 和 DATAFILE_TMP。

[TBLSPACE]:

默认值: None

Oracle 数据库特有参数,执行测试用例时使用的表空间,默认为 'test_' + USER。

[TBLSPACE_TMP]:

默认值：None

Oracle 数据库特有参数，执行测试用例时使用的临时表空间，默认为 'test_' + USER+'_temp'。

[DATAFILE]:

默认值：None

Oracle 数据库特有参数，TBLSPACE 所使用的 datafile 名字，默认为 TBLSPACE + '.dbf'。

[DATAFILE_TMP]:

默认值：None

Oracle 数据库特有参数，TBLSPACE_TMP 所使用的 datafile 名字，默认为 TBLSPACE_TMP + '.dbf'。

[DATAFILE_MAXSIZE]:

默认值：'500M'

Oracle 数据库特有参数，DATAFILE 所允许的最大值。

[DATAFILE_TMP_MAXSIZE]:

默认值：'500M'

Oracle 数据库特有参数，DATAFILE_TMP 所允许的最大值。

[DATAFILE_SIZE]:

默认值：'50M'

Oracle 数据库特有参数，DATAFILE 的初始值。

[DATAFILE_TMP_SIZE]:

默认值：'50M'

Oracle 数据库特有参数，DATAFILE_TMP 的初始值。

[DATAFILE_EXTSIZE]:

默认值：'25M'

Oracle 数据库特有参数，DATAFILE 每次空间不足时所申请的空间大小。

[DATAFILE_TMP_EXTSIZE]:

默认值：'25M'

Oracle 数据库特有参数，DATAFILE_TMP 每次空间不足时所申请的空间大小。

5.3.2　DATABASE_ROUTERS

默认值：[]

数据库路由配置，当执行数据库操作时，Django 会根据路由配置选择恰当的数据库执行操作。默认值是一个空列表：[]，列表元素是一个实现了特殊路由方法的 Python 类的路径。

要想使用数据库路由，首先需要创建数据库路由类，该类必须实现以下方法：

- ❑ db_for_read(model, **hints)

  指定对 model 进行读操作的数据库。

- ❑ db_for_write(model, **hints)

  指定对 model 进行写操作的数据库。

- ❑ allow_relation(obj1, obj2, **hints)

  如果允许 obj1 和 obj2 之间存在关联则返回 True；如果禁止 obj1 和 obj2 之间存在关联则返回

False；如果对 obj1 和 obj2 之间没有限制则返回 None。

这个方法仅用于验证两个对象间是否可以存在外键关联或者多对多关联。

❏ allow_migrate(db, app_label, model_name=None, **hints)

该方法用于确定数据库是否可以进行 migration 操作，第一个参数 db 是数据库的别名。方法返回 True 则表示允许进行 migration 操作；返回 False 则表示不允许进行 migration 操作；返回 None 表示没有特殊规定。

参数 app_label 是进行 migration 操作的应用程序名。

下面举一个数据库路由的例子。假设存在如下数据库配置：

```python
DATABASES = {
    'default': {},
    'auth_db': {
        'NAME': 'auth_db',
        'ENGINE': 'django.db.backends.mysql',
        'USER': 'mysql_user',
        'PASSWORD': 'swordfish',
    },
    'primary': {
        'NAME': 'primary',
        'ENGINE': 'django.db.backends.mysql',
        'USER': 'mysql_user',
        'PASSWORD': 'spam',
    },
    'replica1': {
        'NAME': 'replica1',
        'ENGINE': 'django.db.backends.mysql',
        'USER': 'mysql_user',
        'PASSWORD': 'eggs',
    },
    'replica2': {
        'NAME': 'replica2',
        'ENGINE': 'django.db.backends.mysql',
        'USER': 'mysql_user',
        'PASSWORD': 'bacon',
    },
}
```

下面创建一个用于处理 auth 应用向 auth_db 数据库发送请求的路由类：

```python
class AuthRouter:
    """
    用于处理所有 auth 应用的数据库请求的路由。
    """
    def db_for_read(self, model, **hints):
        """
        尝试在 auth_db 数据库中读取 model。
        """
        if model._meta.app_label == 'auth':
            return 'auth_db'
        return None

    def db_for_write(self, model, **hints):
```

```python
        """
        尝试在auth_db数据库中对model进行写操作。
        """
        if model._meta.app_label == 'auth':
            return 'auth_db'
        return None

    def allow_relation(self, obj1, obj2, **hints):
        """
        允许auth应用中的model存在关系。
        """
        if obj1._meta.app_label == 'auth' or \
           obj2._meta.app_label == 'auth':
            return True
        return None

    def allow_migrate(self, db, app_label, model_name=None, **hints):
        """
        仅允许auth应用中的auth_db进行migration操作。
        """
        if app_label == 'auth':
            return db == 'auth_db'
        return None
```

接下来创建一个路由，用于处理其他所有应用向数据库集群发送的请求。对于读操作，则在集群中随机选择一个数据库：

```python
import random

class PrimaryReplicaRouter:
    def db_for_read(self, model, **hints):
        """
        随机选择集群中的数据库进行读操作。
        """
        return random.choice(['replica1', 'replica2'])

    def db_for_write(self, model, **hints):
        """
        仅使用集群中的主数据库进行写操作。
        """
        return 'primary'

    def allow_relation(self, obj1, obj2, **hints):
        """
        如果两个对象同时存在于数据库集群的所有数据库中，则允许在设置对象间设置关系。
        """
        db_list = ('primary', 'replica1', 'replica2')
        if obj1._state.db in db_list and obj2._state.db in db_list:
            return True
        return None

    def allow_migrate(self, db, app_label, model_name=None, **hints):
        """
```

允许所有非 auth 应用的数据库进行 migration 操作。
"""
return True
```

接下来设置路由信息：

```
DATABASE_ROUTERS = ['AuthRouter 的 Python 路径', 'PrimaryReplicaRouter 的 Python 路径']
```

> 所有数据库操作都会按照路由在 DATABASE_ROUTERS 中出现的顺序进行匹配，在本例中会先使用 AuthRouter 执行数据库操作。
> 此时任何 auth 应用中的数据库操作都将采用路由 AuthRouter，其他应用中的数据库操作都将采用路由 PrimaryReplicaRouter。

除了使用 DATABASE_ROUTERS 外，还可以通过手工指定数据库的方式执行数据库操作，例如：
- Question.objects.using('default').all()
- Question.save(using='default')

5.3.3　DEFAULT_INDEX_TABLESPACE

默认值：''（空字符串）
字段索引的默认表空间。

5.3.4　DEFAULT_TABLESPACE

默认值：''（空字符串）
模型的默认表空间。

5.4　调试

5.4.1　DEBUG

默认值：False
用于指定当前网站是否运行在调试模式下。如果网站已经部署在生产环境，那么一定不能开启 DEBUG 模式。
当开启 DEBUG 模式后，如果网站运行过程中出现异常情况，那么异常信息会被输出到网页上，这些信息包括 Django 执行过程、环境信息、settings.py 中的配置信息。出于安全考虑，Django 不会输出配置中的敏感内容，例如 SECRET_KEY。Django 会根据以下文字对 settings.py 中的信息进行过滤：
- API
- KEY
- PASS
- SECRET
- SIGNATURE
- TOKEN
```

> **注意**
>
> Django 是按照模糊配置的方式在 settings.py 中对以上文字进行匹配的,例如 PASSWORD 会与 PASS 匹配成功。虽然如此,在输出信息中仍然会包含一些敏感信息。

如果 DEBUG 设置为 True,Django 还会记录执行过的每一条 SQL 语句。这对调试来说非常有用,但是会占用大量内存。

最后需要注意的是,当 DEBUG 被设置为 False 时,Django 会认为当前程序已经被部署到生产环境,因此还需要提供可以访问网站的主机信息,相关配置在 ALLOWED_HOSTS 中设置。

下图是 DEBUG=True 时输入的部分错误信息。

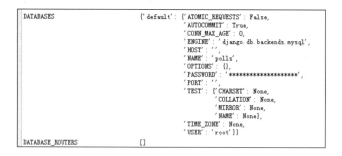

5.4.2 DEBUG_PROPAGATE_EXCEPTIONS

默认值:`False`

如果该参数被设置为 True,那么当 Django 的视图方法出现异常或者需要返回 HTTP 500 时,异常信息将会被 Web 服务器接收并处理而不是由 Django 来处理。

如果希望看到详细异常信息则不应该将它设置为 True。

下图是 DEBUG_PROPAGATE_EXCEPTIONS 为 True 时,网页出现异常时的情况。

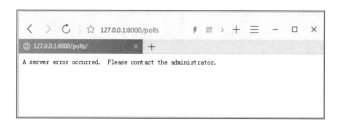

5.5 电子邮件

5.5.1 ADMINS

默认值:`[]`

能够获得代码错误信息的人员列表。当配置文件设置了 DEBUG=False,同时 LOGGING 参数设置了 AdminEmailHandler(默认已设置)时,Django 会将 Web 请求中发生的详细异常信息发送给列表中的所有人。

列表项是一个包含姓名、邮件的元组，例如：

```
[('John', 'john@example.com'), ('Mary', 'mary@example.com')]
```

5.5.2 DEFAULT_FROM_EMAIL

默认值：`'webmaster@localhost'`
默认的电子邮件地址，用于站点管理员发送的各种自动通信。

注意

这些通信不包含向管理员发送异常信息。

5.5.3 EMAIL_BACKEND

默认值：`'django.core.mail.backends.smtp.EmailBackend'`
默认的邮件引擎。

5.5.4 EMAIL_HOST

默认值：`'localhost'`
邮件服务器。

5.5.5 EMAIL_HOST_USER

默认值：`''`（空字符串）
用于登录邮件服务的用户名。

5.5.6 EMAIL_HOST_PASSWORD

默认值：`''`（空字符串）
用于登录邮件服务的用户密码。

5.5.7 EMAIL_PORT

默认值：`25`
SMTP 服务的端口号。

5.5.8 EMAIL_TIMEOUT

默认值：`None`
发送邮件的超时时间，单位为秒。

5.5.9 SERVER_EMAIL

默认值：`'root@localhost'`
用于发送错误信息的邮件地址。

5.5.10 MANAGERS

默认值：`[]`
取值与 ADMINS 一致，用于设置可以接收失效链接的邮箱地址。

5.6 文件上传

5.6.1 DEFAULT_FILE_STORAGE

默认值：`'django.core.files.storage.FileSystemStorage'`
在没有特殊指定文件系统时，任何文件操作所使用的文件存储类。

5.6.2 FILE_CHARSET

默认值：`'utf-8'`
从磁盘读取文件时所使用的编码格式。包括读取模板文件、静态文件和翻译文件。
Django 2.2 版本后已移除，从 Django 3.1 开始所有文件读操作都是 UTF-8 编码。

5.6.3 FILE_UPLOAD_HANDLERS

默认值：

```
[
    'django.core.files.uploadhandler.MemoryFileUploadHandler',
    'django.core.files.uploadhandler.TemporaryFileUploadHandler',
]
```

文件上传处理程序。修改本参数将会改变 Django 文件上传的处理过程。

5.6.4 FILE_UPLOAD_MAX_MEMORY_SIZE

默认值：`2621440`（约 2.5MB）
允许上传的文件最大体积，单位字节。

5.6.5 FILE_UPLOAD_PERMISSIONS

默认值：`0o644`
为新上传的文件设置权限，可选值是 Linux 文件权限的数值模式，如 0o777 表示文件所有者、所在组成员、其他组成员对文件有读、写、执行权限。在大多数操作系统中，临时文件的权限是 0o600。
如果该参数值为 None，那么文件上传时将使用操作系统的默认行为。
出于安全考虑，该参数不会影响 FILE_UPLOAD_TEMP_DIR 所指定的临时文件路径中的文件。

5.6.6 FILE_UPLOAD_DIRECTORY_PERMISSIONS

默认值：`None`
为上传文件过程中所创建的文件夹设置权限，可选值是 Linux 文件权限的数值模式，如 0o777 表示路径所有者、所在组成员、其他组成员对路径有读、写、执行权限。

5.6.7 FILE_UPLOAD_TEMP_DIR

默认值：None

文件上传时的临时存放路径（一般只有当文件大于 FILE_UPLOAD_MAX_MEMORY_SIZE 规定的大小时才会生成临时文件）。如果该参数值为 None，Django 会使用操作系统的默认临时路径，如 /tmp。

5.6.8 MEDIA_ROOT

默认值：''（空字符串）

用于保存用户上传文件的绝对路径，如 /var/www/example.com/media/。

 注意

MEDIA_ROOT 与后面即将介绍的 STATIC_ROOT 的值必须不同。

5.6.9 MEDIA_URL

默认值：''（空字符串）

用于处理 MEDIA_ROOT 中所保存文件的 URL，如果 MEDIA_URL 的值不为空，则必须以斜杠(/) 结尾，例如 http://media.example.com/。

在使用 MEDIA_ROOT 之前需要进行以下配置：

```
from django.conf import settings
from django.conf.urls.static import static

urlpatterns = [
    # 其他 URL 配置代码
] + static(settings.MEDIA_URL, document_root=settings.MEDIA_ROOT)
```

如果希望在模板中使用 {{ MEDIA_URL }}，需要将 django.template.context_processors.media 添加到 TEMPLATES 配置中，如：

```
TEMPLATES = [
    {
        'BACKEND' : 'django.template.backends.django.DjangoTemplates',
        'DIRS': [os.path.join(BASE_DIR, 'templates'),
            os.path.join(os.path.dirname(os.path.abspath(__file__)), 'templates')],
        'APP_DIRS': True,
        'OPTIONS': {
            'context_processors': [
                'django.template.context_processors.debug',
                'django.template.context_processors.request',
                'django.contrib.auth.context_processors.auth',
                'django.contrib.messages.context_processors.messages',
                'django.template.context_processors.media',
            ],
        },
    },
]
```

5.6.10 静态文件

1. STATIC_ROOT

默认值：None

使用 django-admin collectstatic 命令收集静态文件时，静态文件归集的文件夹。当执行 collectstatic 命令时，项目中的所有静态文件将被复制到 STATIC_ROOT 所指定的路径，包括 admin 项目中的静态文件。

2. STATIC_URL

默认值：None

引用 STATIC_ROOT 中的静态文件时所使用的 URL。

例如："/static/" 或者 http://static.example.com/。

不为 None 时，该参数将用于确认资源文件的基础路径。设置该参数时必须以反斜杠 (\) 结尾。

3. STATICFILES_DIRS

默认值：[]

除了默认的静态文件路径（Django 项目的 static 文件夹）外，额外的静态文件路径。collectstatic 命令会同时收集这些路径下的静态文件。

例如：

```
STATICFILES_DIRS = [
    "/home/special.polls.com/polls/static",
    "/home/polls.com/polls/static",
    "/opt/webfiles/common",
]
```

需要注意的是，这里的路径要按照 UNIX 风格书写，路径中的分隔符一律用斜杠 (/)。

为了使用方便，还可以给静态文件路径添加命名空间，例如：

```
STATICFILES_DIRS = [
    # ...
    ("downloads", "/opt/webfiles/stats"),
]
```

假设 STATIC_URL='/static/'，此时可以使用以下方式引用静态文件：

```
'/static/downloads/polls_20101022.tar.gz'
```

在模板中可以用以下方式引用静态文件：

```
<a href="{% static "downloads/polls_20101022.tar.gz" %}">
```

4. STATICFILES_STORAGE

默认值：'django.contrib.staticfiles.storage.StaticFilesStorage'

使用 collectstatic 命令收集静态文件时所使用的文件存储引擎。

5. STATICFILES_FINDERS

默认值：

```
[
    'django.contrib.staticfiles.finders.FileSystemFinder',
```

```
        'django.contrib.staticfiles.finders.AppDirectoriesFinder',
]
```

用于查找静态文件的文件检索引擎。FileSystemFinder 用于查找 STATICFILES_DIRS 配置项下面的文件，AppDirectoriesFinder 用于查找每个应用程序的 static 目录。如果出现多个重名文件，则使用第一个查找到的文件。

Django 默认禁用了一个文件检索引擎：django.contrib.staticfiles.finders.DefaultStorageFinder，该引擎会查找 DEFAULT_FILE_STORAGE 配置项下的静态文件。

5.7 表单

5.7.1 FORM_RENDERER

默认值：`'django.forms.renderers.DjangoTemplates'`
用于渲染表单挂件的 Python 类，必须实现底层渲染 API。

5.8 国际化（i18n/l10n）

5.8.1 DECIMAL_SEPARATOR

默认值：`'.'`
格式化十进制数时所使用的默认分隔符。
需要注意的是，如果将 USE_L10N 设置为 True，那么本地语言格式将会有更高的优先级。
例如，当 DECIMAL_SEPARATOR = '*' 时，使用 floatformat 格式化小数时将使用 '*' 替代默认的句点。

模板代码：`<span>{{ 123.456789|floatformat }}</span>`
浏览器显示样式：123*5

5.8.2 NUMBER_GROUPING

默认值：0
数字整数部分的分组规则，通常使用千分位分割符。如果设置为 0，那么数字不会分组显示；如果大于 0，就会使用 THOUSAND_SEPARATOR 对数字进行分组。

某些特殊地区的语言使用非标准的数字分组方式，他们的数字每一组所包含的长度不相同，例如 en_IN：10,00,00,000。对于这种情况需要给出具体数字分组所需的数字长度，元组中第一个数字代表小数点前第一组的长度，元组中第二个数字代表小数点前第二组的长度，依次类推，如果实际数字长度大于分组数，那么元组中的最后一个数字代表实际数字后续分组的长度。如果元组中最后一个数字是 -1，那么将终止数字的后续分组。如果元组中最后一个数字是 0，那么最后一个分组长度将应用于剩余的全部数字。

代码示例 1
配置文件：
```
LANGUAGE_CODE = 'ta'
```

```
USE_I18N = False
USE_L10N = True
USE_TZ = True
NUMBER_GROUPING = 2
THOUSAND_SEPARATOR = ' '
USE_THOUSAND_SEPARATOR = True
```

模板代码：

```
{{ 123456789012456.012345|intcomma }}
```

显示样式：

```
1 23 45 67 89 01 24 56.02
```

代码示例 2

配置文件：

```
LANGUAGE_CODE = 'ta'
USE_I18N = False
USE_L10N = True
USE_TZ = True
NUMBER_GROUPING = (3,2,0)
THOUSAND_SEPARATOR = ' '
USE_THOUSAND_SEPARATOR = True
```

模板代码：

```
{{ 123456789012456.012345|intcomma }}
```

显示样式：

```
12 34 56 78 90 12 456.02
```

代码示例 3

配置文件：

```
LANGUAGE_CODE = 'ta'
USE_I18N = False
USE_L10N = True
USE_TZ = True
NUMBER_GROUPING = (3,2,-1)
THOUSAND_SEPARATOR = ' '
USE_THOUSAND_SEPARATOR = True
```

模板代码：

```
{{ 123456789012456.012345|intcomma }}
```

显示样式：

```
1234567890 12 456.02
```

代码示例 4

配置文件：

```
LANGUAGE_CODE = 'en-us'
USE_I18N = False
USE_L10N = True
USE_TZ = True
```

```
NUMBER_GROUPING = (3,2,-1)
THOUSAND_SEPARATOR = ' '
USE_THOUSAND_SEPARATOR = True
```

模板代码：

```
{{ 123456789012456.012345|intcomma }}
```

显示样式：

```
123,456,789,012,456.02
```

代码分析：

细心的读者可能发现代码示例 3 和代码示例 4 除了所使用的语言（LANGUAGE_CODE）不同外，其他代码完全相同，但是输出结果却相去甚远。出现这种情况的原因是当 USE_L10N 被设置为 True 时，Django 会使用预置的本地化文件来格式化数字，Django 的本地化文件存储在 django/conf/locale/ 下，例如 en 对应的本地化文件是 django/conf/locale/en/。此时即使在 settings.py 中设置了数字的显示样式，但是仍然按照 Django 默认的本地化格式转换数字样式。但是为什么 LANGUAGE_CODE = 'ta' 时能够按照 settings.py 中的设置进行显示呢？这是由于随着 Django 版本一共发布了 90 多个地区的本地化文件（这里指 Django 2.2 版本，后续还会增加），只有以下 5 个语言没有本地化：

./ta/formats.py

./fy/formats.py

./te/formats.py

./mn/formats.py

./kn/formats.py

按照 Django 官方的说法，如果完全禁用本地化（USE_L10N = False），就不会出现上面的问题了。但是，如果完全禁用本地化，又会导致真正需要本地化的地方出现错误，而实际应用中完全禁用本地化也不能解决数字的本地化显示问题，例如 LANGUAGE_CODE = 'ta' 将不能使用 settings.py 中的设置。

配置文件：

```
LANGUAGE_CODE = 'ta'
USE_I18N = False
USE_L10N = False
USE_TZ = True
NUMBER_GROUPING = (3,2,-1)
THOUSAND_SEPARATOR = ' '
USE_THOUSAND_SEPARATOR = True
```

模板代码：

```
{{ 123456789012456.012345|intcomma }}
```

显示样式：

```
123,456,789,012,456.02
```

解决以上问题的最好办法是编写自定义模板过滤器，这将在第 10 章中详细介绍。

5.8.3 THOUSAND_SEPARATOR

默认值：','

格式化数字时所使用的千分位分隔符。只有当 USE_THOUSAND_SEPARATOR=True 且 NUMBER_

GROUPING 大于 0 时该参数才生效。

如果 USE_L10N 设置为 True，那么本地语言格式将会有更高的优先级。

5.8.4　USE_THOUSAND_SEPARATOR

默认值：`False`

标志是否使用千分位分隔符显示数字类型。当 USE_THOUSAND_SEPARATOR=True 且 USE_L10N=True 时，Django 将会使用 THOUSAND_SEPARATOR 和 NUMBER_GROUPING 来格式化数字类型，除非当前地区（locale）已经有了千分位分隔符。当前地区的千分位分隔符享有最高优先级，这就是 NUMBER_GROUPING 中讨论的问题的根本原因。

5.8.5　FIRST_DAY_OF_WEEK

默认值：`0`（星期日）

表示每周第一天的数字。该参数在显示日历时非常有用，参数值必须是 0～6 的整数，0 代表星期日，1 代表星期一。

5.8.6　DATE_FORMAT

默认值：`'N j, Y'`

显示效果：Feb. 4, 2003。

系统中日期字段的默认显示格式。需要注意的是，如果 USE_L10N 设置为 True，那么本地语言环境的显示样式将拥有更高的优先级。中文情况下日期的显示格式可能是这样的：2003 年 2 月 4 日。

5.8.7　DATE_INPUT_FORMATS

默认值：

```
[
    '%Y-%m-%d', '%m/%d/%Y', '%m/%d/%y',  # '2006-10-25', '10/25/2006', '10/25/06'
    '%b %d %Y', '%b %d, %Y',              # 'Oct 25 2006', 'Oct 25, 2006'
    '%d %b %Y', '%d %b, %Y',              # '25 Oct 2006', '25 Oct, 2006'
    '%B %d %Y', '%B %d, %Y',              # 'October 25 2006', 'October 25, 2006'
    '%d %B %Y', '%d %B, %Y',              # '25 October 2006', '25 October, 2006'
]
```

一组在日期字段中输入日期时所使用的格式。Django 会按照时间格式在列表中的顺序依次对比，直到找到第一个匹配的时间格式。需要注意的是，这些格式字符串使用的是 Python datetime 模块语法而不是 date 模板过滤器的格式字符串。

如果不按照规定的日期格式输入日期将会出错，如下图所示。

如果 USE_L10N 设置为 True，那么本地语言环境的显示样式将拥有更高的优先级。

5.8.8　DATETIME_FORMAT

默认值：`'N j, Y, P'`

显示效果：Feb. 4, 2003, 4 p.m.。

系统中日期时间字段的默认显示格式。需要注意的是，如果 USE_L10N 设置为 True，那么本地语言环境的显示样式将拥有更高的优先级。中文情况下日期的显示格式可能是这样的：2019 年 9 月 20 日 15:31。

5.8.9　SHORT_DATE_FORMAT

默认值：`'m/d/y'`

显示效果：10/10/2019。

模板中日期字段的显示样式。

如果 USE_L10N 设置为 True，那么本地语言环境的显示样式将拥有更高的优先级。

5.8.10　SHORT_DATETIME_FORMAT

默认值：`'m/d/y P'`

显示效果：10/10/2019 1PM。

模板中日期时间字段的显示样式。

如果 USE_L10N 设置为 True，那么本地语言环境的显示样式将拥有更高的优先级。

5.8.11　DATETIME_INPUT_FORMATS

默认值：

```
[
    '%Y-%m-%d %H:%M:%S',      # '2006-10-25 14:30:59'
    '%Y-%m-%d %H:%M:%S.%f',   # '2006-10-25 14:30:59.000200'
    '%Y-%m-%d %H:%M',         # '2006-10-25 14:30'
    '%Y-%m-%d',               # '2006-10-25'
    '%m/%d/%Y %H:%M:%S',      # '10/25/2006 14:30:59'
    '%m/%d/%Y %H:%M:%S.%f',   # '10/25/2006 14:30:59.000200'
    '%m/%d/%Y %H:%M',         # '10/25/2006 14:30'
    '%m/%d/%Y',               # '10/25/2006'
    '%m/%d/%y %H:%M:%S',      # '10/25/06 14:30:59'
    '%m/%d/%y %H:%M:%S.%f',   # '10/25/06 14:30:59.000200'
    '%m/%d/%y %H:%M',         # '10/25/06 14:30'
    '%m/%d/%y',               # '10/25/06'
]
```

一组在日期时间字段中输入日期时间时所使用的格式。Django 会按照时间格式在列表中的顺序依次对比，直到找到第一个匹配的时间格式。需要注意的是，这些格式字符串使用的是 Python datetime 模块语法而不是 date 模板过滤器的格式字符串。如果 USE_L10N 设置为 True，那么本地语言环境的显示样式将拥有更高的优先级。

如果不按照规定的日期格式输入日期将会出错，如下图所示。

5.8.12 TIME_FORMAT

默认值：`'P'`

显示效果：(e.g. 4 p.m.)。

系统中时间字段的默认显示格式。需要注意的是，如果 USE_L10N 设置为 True，那么本地语言环境的显示样式将拥有更高的优先级。

5.8.13 TIME_INPUT_FORMATS

默认值：

```
[
    '%H:%M:%S',            # '14:30:59'
    '%H:%M:%S.%f',         # '14:30:59.000200'
    '%H:%M',               # '14:30'
]
```

一组在时间字段中输入时间时所使用的格式。Django 会按照时间格式在列表中的顺序依次对比，直到找到第一个匹配的时间格式。需要注意的是，这些格式字符串使用的是 Python datetime 模块语法而不是 date 模板过滤器的格式字符串。如果 USE_L10N 设置为 True，那么本地语言环境的显示样式将拥有更高的优先级。

5.8.14 YEAR_MONTH_FORMAT

默认值：`'F Y'`

Django admin 页面上变更列表中日期字段所使用的格式。如果其他任何地方仅显示年和月，也有可能会用到当前格式。

不同地区的显示格式也是不同的，例如，U.S. English 可能显示 "January 2006"，但是其他地区可能显示 "2006/January"。

5.8.15 MONTH_DAY_FORMAT

默认值：`'F j'`

Django admin 页面上变更列表中日期字段所使用的格式。如果其他任何地方仅显示年和月，也有可能会用到当前格式。

需要注意的是，如果 USE_L10N 设置为 True，那么对应的本地语言格式将享有更高的优先级。

5.8.16 TIME_ZONE

默认值：`'America/Chicago'`

因为 Django 第一个发布版本的时区是 `'America/Chicago'`，所以 `'America/Chicago'` 就一直是 Django 项目的默认全局配置。新的项目模板默认使用 `'UTC'` 时区。

需要注意的是，Django 项目的时区并非一定要与服务器的时区相同，有时一台服务器上部署了多个 Django 项目，每个项目可以有不同的时区。

如果 USE_TZ 设置为 False，那么 Django 会使用这个参数所指定的时区来保存全部日期时间；如果 USE_TZ 设置为 True，那么 TIME_ZONE 是 Django 模板中显示日期时间的默认时区。

需要注意的是，Django 官方文档提到在 Windows 系统上 Django 不能准确使用其他不同于本地时区的时区，因此在 Windows 系统上只能将 TIME_ZONE 设置为本地时区。但是，笔者在 Windows 10 系统上使用时还是能够正常显示和保存芝加哥时间的，如下图所示。

出于对系统可靠性的考虑，如果在 Windows 系统上运行 Django 项目，最好使用系统时区，除非你已经验证了在你的项目中所有用到时区的地方都能够正常工作。

中国所在时区为 Asia/Shanghai，如果将时区设置为 Asia/Chongqing、Asia/Harbin 等中国时区，其本质就是 Asia/Shanghai。

所有可选时区可以在维基百科中查到：

https://en.wikipedia.org/wiki/List_of_tz_database_time_zones

5.8.17 LANGUAGE_CODE

默认值：`'en-us'`
当前程序所使用的语言。简体中文为 `'zh-Hans'`。
为了使本参数生效必须将 USE_I18N 设置为 True。
所有可选语言编码可以在 i18nguy 查看：

http://www.i18nguy.com/unicode/language-identifiers.html

5.8.18 LANGUAGE_COOKIE_AGE

默认值：`None`
Language Cookie 的有效期，单位为秒。如果设置为 None，那么有效期是浏览器的关闭时间。

5.8.19 LANGUAGE_COOKIE_DOMAIN

默认值：`None`
Language Cookie 所使用的域。如果需要跨域 Cookie（cross-domain Cookie），那么就将该参数设

置为对应的域名，例如 example.com。否则将使用标准的域 Cookie（standard domain Cookie）。

在生产环境上变更该参数时一定要小心。如果将一个使用标准域 Cookie 的网站修改为使用跨域 Cookie，那么那些已经存在的用户仍然会使用原有的 Cookie。这会导致这些用户不能够切换语言。处理这个问题唯一安全可靠的方法是更改 Cookie 的名字（通过设置 LANGUAGE_COOKIE_NAME 完成），同时编写方法复制现有的用户 Cookie 到新 Cookie，然后删除旧 Cookie。

5.8.20　LANGUAGE_COOKIE_NAME

默认值：`'django_language'`

Language Cookie 的名字。可以任意设置，只要不与其他 Cookie 重名就行。

5.8.21　LANGUAGE_COOKIE_PATH

默认值：`'/'`

Language Cookie 的路径。可以是 URL 路径也可以是 URL 的父级路径。通过设置 Language Cookie 的路径，可以使同一个服务器上的不同 Django 站点互相隔离。

在生产环境更新 Language Cookie 的路径时同样要非常小心，因为已有的用户数据不会被更新。解决办法是使用 LANGUAGE_COOKIE_NAME 新建 Cookie，然后更新已有 Cookie，并且删除旧 Cookie。

5.8.22　LANGUAGES

默认值：一组可用的语言。

随着 Django 版本的迭代，这个列表也在不断增长，当前版本所支持的语言可以在全局配置文件中查看（django/conf/global_settings.py）。

这个数组的每一个元素都是一个元组，元组包含两个元素（语言码和语言名），例如（'ja', 'Japanese'）。这个数组决定了可以在国际化和本地化中选择哪些语言。

通常情况下默认值就已经够用，设置该参数的主要目的是约束 Django 可选择的语言。

5.8.23　LANGUAGES_BIDI

默认值：一组使用 BiDi（从右至左）格式书写的语言的语言码。

5.8.24　LOCALE_PATHS

默认值：`[]`

一组 Django 查找翻译文件的路径。

代码示例：

```
LOCALE_PATHS = [
    '/home/www/project/common_files/locale',
    '/var/local/translations/locale',
]
```

5.8.25　USE_I18N

默认值：True

规定 Django 是否启用翻译系统，若设置为 True，Django 会将系统语言翻译成 LANGUAGE_

5.8.26 USE_L10N

默认值：`False`

规定 Django 是否启用本地化系统，若设置为 True，Django 会对数字、日期、时间字符串进行格式化。默认值为 False。

5.8.27 USE_TZ

默认值：`False`

规定 datetime 类型是否根据时区进行显示。如果 USE_TZ 设置为 True，Django 内部的所有 datetime 类型将根据不同时区进行处理。

例如，当 USE_TZ 设置为 True 时，在 Admin 后台管理系统保存一条调查问卷，检查数据库中保存的时间以及网页中显示的时间。

数据库中保存的是 UTC 时间，如下图所示。

网页中显示的是本地时间：

5.8.28 Python datetime 语法

Python 提供了一个格式化日期时间的方法：strftime(format)。通过格式化字符串可以对 date、datetime、time 对象进行格式化，从而得到一个新的字符串。这个格式化字符串就是 Django 中日期时间配置中所使用的格式化字符串，具体使用方法可参考 Python 官方文档：

https://docs.python.org/3/library/datetime.html#strftime-strptime-behavior

下面举个例子来说明 Python 的日期时间格式化方法：

```
>>> today = datetime.datetime.now()
>>> today.strftime("%Y %B %d %A")
'2019 September 30 Monday'
```

5.9 HTTP

5.9.1 DATA_UPLOAD_MAX_MEMORY_SIZE

默认值：`2621440`（2.5MB）

每次请求所允许的最大字节数，超过这个数值会抛出 SuspiciousOperation 异常。当访问 request.

body 或者 request.POST 的时候会进行检查，检查会计算除了上传的文件外的请求大小。可以通过设置 None 来禁用请求大小的检查。

由于请求数据的大小直接影响到 GET、POST 等方法占用内存的大小，因此大的请求经常被用于拒绝服务攻击。

5.9.2　DATA_UPLOAD_MAX_NUMBER_FIELDS

默认值：`1000`

一次 GET 或者 POST 请求所允许传递的最大参数数量，超出则报 SuspiciousOperation 异常。可以通过设置 None 来禁用请求大小的检查。

请求参数的数量会影响请求执行的时间以及从 GET 或者 POST 字典中提取字段的效率。拒绝服务攻击常用的手段就是向服务器发送较大数量的请求参数。

5.9.3　DEFAULT_CHARSET

默认值：`'utf-8'`

如果在 HTML 代码中没有使用 MIME 指定字符集，那么 DEFAULT_CHARSET 将会用于所有 HttpResponse 对象。该参数与 DEFAULT_CONTENT_TYPE 共同构建了 HTML 头部中的 Content-Type。

5.9.4　DISALLOWED_USER_AGENTS

默认值：`[]`

一组已经编译的用于匹配 User-Agent 字符串的正则表达式对象，当请求的 User-Agent 与这些正则表达式中的任意一个匹配成功时将不能访问网站的任意页面。使用该参数需要保证已经安装了 CommonMiddleware 中间件。

该参数主要用于防止互联网机器人爬虫。

参数示例：

```
import re

DISALLOWED_USER_AGENTS = (
  re.compile(r'^OmniExplorer_Bot'),
  re.compile(r'^Googlebot')
)
```

5.9.5　FORCE_SCRIPT_NAME

默认值：`None`

如果该参数值不为 None，那么参数值将作为任何 HTTP 请求的 SCRIPT_NAME 环境变量。

5.9.6　INTERNAL_IPS

默认值：`[]`

一组 IP 地址字符串，用于以下场景：
- 允许 debug() 方法向模板中添加变量。

- 非注册用户使用管理员页面。
- 在 AdminEmailHandler 邮件中标注为 internal（与之对应的是 EXTERNAL）。

5.9.7　SECURE_BROWSER_XSS_FILTER

默认值：`False`

如果设置为 True，SecurityMiddleware 将会在每个 HTTP 响应头添加 X-XSS-Protection: 1; mode=block。如果浏览器具有阻止 XSS 攻击的能力，此时浏览器将会阻止一切可疑的 XSS 攻击脚本。

5.9.8　SECURE_CONTENT_TYPE_NOSNIFF

默认值：`True`

如果设置为 True，SecurityMiddleware 将会在每个 HTTP 响应头添加 X-Content-Type-Options: nosniff。

某些浏览器会尝试猜测网页内容的类型，从而重写 Content-Type。当服务器端设置与客户端不匹配的情况下，这种重写 Content-Type 的操作会有一定的帮助，但是也会带来一定的安全隐患。

试想如果你的 Web 站点提供文件上传下载服务，此时攻击者上传一个精心制作的 HTML 或者 JavaScript 文件，虽然你希望浏览器能够原样显示这些文件，但是浏览器将这些文件识别成对应的 HTML 或者 JavaScript 文件，从而被非法执行，这将会带来严重的安全隐患。

为了防止发生以上行为，开发人员应该阻止浏览器猜测文件的类型并严格按照服务器端发送的 Content-Type 解释文件，此时只需要设置 SECURE_CONTENT_TYPE_NOSNIFF 为 True 即可。

注意

如果 Django 应用并不直接处理文件上传下载，那么就不需要设置该参数。

5.9.9　SECURE_HSTS_INCLUDE_SUBDOMAINS

默认值：`False`

如果设置为 True，SecurityMiddleware 将会在 HTTP Strict Transport Security 响应头添加 includeSubDomains 指令，前提是需要将 SECURE_HSTS_SECONDS 设置为一个非 0 值。

如果本参数设置错误，可能会导致网站出现不可逆的损坏，需要谨慎使用。

5.9.10　SECURE_HSTS_PRELOAD

默认值：`False`

如果设置为 True，SecurityMiddleware 将会在 HTTP Strict Transport Security 响应头添加 preload 指令，前提是需要将 SECURE_HSTS_SECONDS 设置为一个非 0 值。

5.9.11　SECURE_HSTS_SECONDS

默认值：`0`

如果将本参数设置为一个非 0 值，SecurityMiddleware 将会为所有响应添加 HTTP Strict Transport Security 头。

如果本参数设置错误，可能会导致网站出现不可逆的损坏，需要谨慎使用。

> **小技巧**
>
> 刚刚设置 HSTS 的时候最好先使用使用足够小的值进行测试，如设置一个小时：SECURE_HSTS_SECONDS = 3600，在这一个小时里尝试使用各种方式访问网站，包括使用不安全链接（HTTP）进行访问，当确认网站已经能够正常运行了再将 HSTS 设置为一个足够大的值，如 31 536 000，该值近似一年。

5.9.12　SECURE_PROXY_SSL_HEADER

默认值：`None`

一个 HTTP 头和值的组合，用于表示请求是安全的。该参数决定了 request 对象的 is_secure() 方法。

默认情况下，request 对象的 is_secure() 方法通过判断请求的 URL 是不是 https:// 来决定请求是否安全。这个方法对于 CSRF 保护很重要，同时也能在其他场景中用于安全验证。

但是，如果你的 Django 应用程序位于代理的后面，那么无论原始请求是否使用 HTTPS，代理都可能会执行。如果在代理和 Django 应用之间没有使用 https 链接，is_secure() 方法将会返回 False，反之将永远返回 True。

在这种情况下就可以通过设置 SECURE_PROXY_SSL_HEADER 来告诉 Django 当前请求是安全的。SECURE_PROXY_SSL_HEADER 参数值是一个包含两个元素的元组，第一个元素是 HTTP 请求头，第二个元素是 HTTP 请求头对应的值，例如：

```
SECURE_PROXY_SSL_HEADER = ('HTTP_X_FORWARDED_PROTO', 'https')
```

上面的设置告诉 Django 要信任包含 X_Forwarded_Proto 头的所有请求，因为这些请求是我们自己的代理服务器发送过来的，元组的第二个元素固定为 'https'。

> **注意**
>
> 只有当你的 Django 应用程序部署在代理上才需要配置该参数，其他情况除非你明确需要使用该参数否则都不要设置该参数。

由于 Django 会自动在 x-header 前添加 'HTTP_'，因此，在设置该参数时一定要按照 request.META 格式书写（全部大写并在必要时增加 'HTTP_'）。

设置本参数时需要保证以下条件全部成立：

- Django 应用程序部署在代理之后。
- 不要自己编写 X_Forwarded_Proto 请求头，因为代理服务器会从请求中删除 X_Forwarded_Proto 请求头。
- 当原始请求来自于 HTTPS 时，代理设置新的 X_Forwarded_Proto 请求头并发送给 Django。

5.9.13　SECURE_REDIRECT_EXEMPT

默认值：`[]`

如果 URL 路径与此列表中的正则表达式匹配，则请求将不会重定向到 HTTPS。

只有当 SECURE_SSL_REDIRECT 不是 False 时本参数才生效。

5.9.14　SECURE_REFERRER_POLICY

默认值：None

如果配置了该参数，SecurityMiddleware 中间件就会为所有还没有设置 Referrer 请求头的请求设置 Referrer 头，Referrer 头的值是该参数值。

5.9.15　SECURE_SSL_HOST

默认值：None

如果将该参数设置为一个字符串（如 secure.example.com），所有 SSL 重定向将会指向该主机而不是原始主机（如 www.example.com）。

只有当 SECURE_SSL_REDIRECT 不是 False 时本参数才生效。

5.9.16　SECURE_SSL_REDIRECT

默认值：False

如果设置为 True，中间件 SecurityMiddleware 将会把所有非 HTTPS 请求重定向到 HTTPS（除了 SECURE_REDIRECT_EXEMPT 中的 URL）。

如果站点同时支持 HTTP 和 HTTPS 访问，那么大多数用户都会使用 HTTP 去访问你的网站。所有此时最好将 HTTP 请求重定向到 HTTPS。

出于性能方面的考虑，重定向工作最好是放在 Django 之外进行，如使用 nginx。

5.9.17　SIGNING_BACKEND

默认值：'django.core.signing.TimestampSigner'

用于为 Cookie 或其他数据进行签名的后台程序。

5.9.18　WSGI_APPLICATION

默认值：None

指向 WSGI 应用程序的 Python 路径，Django 的内置服务器将会使用这个应用程序运行 Django 代码。

在使用 django-admin startproject 命令创建 Django 工程的时候会自动创建一个 wsgi.py 脚本，在这个脚本中存在一个可执行的 application 对象，WSGI_APPLICATION 就是这个 application 对象的路径。

如果在配置文件中没有设置这个参数，将会使用 django.core.wsgi.get_wsgi_application() 的返回值来指定 WSGI_APPLICATION，runserver 命令不变。

5.10　安全

5.10.1　SECRET_KEY

默认值：''（空字符串）

Django 工程的密钥。该密钥将会用于加密签名，它的值是一个随机字符串，每个工程都不一样。

在使用 django-admin startproject 命令创建工程时会自动生成一个随机密钥。

在项目中不要直接使用 SECURET_KEY，如果一定要用，需要使用 django.utils.encoding.force_text() 或者 django.utils.encoding.force_bytes() 进行安装转换。

如果 SECURET_KEY 是空，Django 将拒绝启动应用程序。

SECURET_KEY 的用途：

- 在除了默认的 session 引擎外的其他引擎中使用，或者在 get_session_auth_hash() 方法中使用。
- 当使用 CookieStorage 和 FallbackStorage 时，在所有消息中使用。
- 在 PasswordResetView 视图的 token 中使用。
- 在加密签名中使用。

5.10.2 ALLOWED_HOSTS

默认值：[]

Django 网站可以提供服务的一组主机或者域名。设置 ALLOWED_HOSTS 是为了防止伪造近似的主机名进行 HTTP 攻击。

数组的值可以是完全限定名（fully qualified names），例如 www.example.com，此时请求头的 Host 值就要求完全匹配 ALLOWED_HOSTS。可以使用句点来匹配全部子域名，例如，".example.com" 可以匹配 example.com、www.example.com 以及任何 example.com 的子域名。还可以使用星号"*"来匹配任意字符，如果这样做，就需要自己完成请求头 Host 的校验了。

除了完全限定名，Django 还支持全限定域名（Fully Qualified Domain Name）。

当请求的 Host 属性值与列表中所有值都不匹配时，执行 django.http.HttpRequest.get_host() 方法会抛出一个 SuspiciousOperation 异常。

如果 DEBUG 参数是 True，同时 ALLOWED_HOSTS 是空数组，ALLOWED_HOSTS 等价于 ['localhost', '127.0.0.1', '[::1]']。

5.11 CSRF

通常黑客可以在其他网站中制造虚假链接诱使用户点击该链接向目标网站发送请求，如果此时用户正在登录目标网站或者用户 session 还没过期，那么目标网站会接收并执行请求。为了避免服务器程序执行这些虚假链接，Django 提供了 CSRF（Cross Site Request Forgery，跨站点请求伪造）验证。本节介绍如何设置 CSRF。

5.11.1 CSRF_COOKIE_AGE

默认值：31449600（单位为秒，接近 1 年）

CSRF Cookie 的有效期。设置超长 CSRF Cookie 有效期的目的是为了防止用户关闭浏览器或者设置了浏览器书签，然后从缓存重新打开浏览器时出现问题。如果没有超长缓存，这时提交表单可能会失败。

由于某些浏览器可以禁用持久 Cookie 或者破坏磁盘上的 Cookie，这会导致 CSRF 检查失败，此时可以将 CSRF_COOKIE_AGE 设置为 None，使用基于 session 的 CSRF Cookie，CSRF Cookie 保存在内存中。

5.11.2 CSRF_COOKIE_DOMAIN

默认值：`None`

设置 CSRF Cookie 时所使用的域名。这个设置可以很方便地将跨子域请求与跨站点请求区分开。参数值应该是一个域名，例如 example.com，此时 a.example.com 的 POST 请求就可以正常地提交给 b.example.com 的视图了。

5.11.3 CSRF_COOKIE_HTTPONLY

默认值：`False`

表示是否在 CSRF Cookie 中使用 HttpOnly。基于 HttpOnly 的特性，如果 CSRF_COOKIE_HTTPONLY 设置为 True，那么客户端的 JavaScript 脚本将不能访问 CSRF Cookie。

需要注意的是，由于 Django 的 CSRF 策略只用于防止跨站请求伪造攻击，因此即使将 CSRF_COOKIE_HTTPONLY 设置为 True，也不能为 CSRF Cookie 提供任何额外的保护。另外，如果攻击者能够通过 JavaScript 读取 Cookie，实际上他就能够做任何想做的事情，因为浏览器会认为攻击者已经登录了正确的网站，即使不让他读取 CSRF Cookie 也没什么用。有时公司领导要求为 CSRF Cookie 添加 HttpOnly，那就添加吧。

需要注意的是，如果启用了 CSRF_COOKIE_HTTPONLY，那么通过 AJAX 提交 CSRF token 的时候不能直接读取 Cookie，需要使用一个隐藏的表单控件。

5.11.4 CSRF_COOKIE_NAME

默认值：`'csrftoken'`

CSRF 认证中所使用的 Cookie 名。可以是任何字符串，只要与其他 Cookie 区别开即可。

5.11.5 CSRF_COOKIE_PATH

默认值：`'/'`

CSRF Cookie 的路径。这个路径通常与 Django 的 URL 或者上一级 URL 相匹配。

这在同一主机下运行多个 Django 实例的时候非常有用，它可以保证每个 Django 实例都有自己的 CSRF Cookie，互不影响。

5.11.6 CSRF_COOKIE_SAMESITE

默认值：`'Lax'`

SameSite 可以阻止浏览器将 Cookie 与跨站点请求一起发送。主要目的是降低跨域信息泄露的风险。它还提供了一些针对跨站点请求伪造攻击的保护。

该标志的可能值为 Lax 或 Strict。读者可以去网络搜索 SameSite 的详细用法，在此只要知道如果 Cookie 的 SameSite 属性被设置为 Strict，那么在进行跨站请求时 Cookie 将不会随网络请求发送出去。

5.11.7 CSRF_COOKIE_SECURE

默认值：`False`

是否为 CSRF Cookie 设置 secure 属性。如果设置为 True，那么这个 Cookie 将只会被 HTTPS 请求发送。

5.11.8　CSRF_USE_SESSIONS

默认值：`False`

是否将 CSRF token 保存在用户会话中而不是保存在 Cookie 中，CSRF_USE_SESSIONS 需要用到 django.contrib.sessions。

虽然将 CSRF token 保存在 Cookie 中已经足够安全，但是有时公司的安全管理部门仍然规定使用 session，此时可以将 CSRF_USE_SESSIONS 设置为 True。

由于默认的异常处理视图需要用到 CSRF token，因此在 MIDDLEWARE 列表中 SessionMiddleware 必须放在第一位。

5.11.9　CSRF_FAILURE_VIEW

默认值：`'django.views.csrf.csrf_failure'`

当请求被 CSRF 保护机制拒绝时所调用的视图。该视图方法签名如下：

```
def csrf_failure(request, reason=""):
    ...
```

参数 reason 是一个对于异常信息的简短描述，主要为开发人员调试使用，返回 HttpResponseForbidden 实例。

5.11.10　CSRF_HEADER_NAME

默认值：`'HTTP_X_CSRFTOKEN'`

用于 CSRF 认证的请求头的名字。与其他保存在 request.META 中的 HTTP 头一样，从服务器端获取的 CSRF_HEADER_NAME 中的全部字符会被自动转换为大写字母、将连字符（-）转换成下画线（_）同时额外增加"HTTP_"前缀，在前端使用 CSRF_HEADER_NAME 的时候要转换一下，例如前端使用 X-CSRFTOKEN，那么配置文件中就要写成 HTTP_X_CSRFTOKEN。

5.11.11　CSRF_TRUSTED_ORIGINS

默认值：`[]`

为不安全请求（例如 POST 请求）提供可信任的主机列表。为了保护不安全的请求，Django 的 CSRF 保护机制要求请求具有一个 Referer 头且该 Referer 应与原始请求的 Host 头匹配。这种机制阻止了不同子域之间相互访问，例如来自 subdomain.example.com 的 POST 请求不能访问 api.example.com。为了解决该问题，可以将 subdomain.example.com 添加到 CSRF_TRUSTED_ORIGINS 中。如果域名 example.com 下存在多个子域名且需要允许所有子域名能够互相访问，可以在列表中添加".example.com"。

5.11.12　代码示例

前面讲了这么多 CSRF 相关的设置，下面通过一个示例进行总结。

首先在 settings.py 中添加一些配置项：

```
CSRF_HEADER_NAME = 'HTTP_X_CSRF_AARONTOKEN'
CSRF_COOKIE_SAMESITE = 'Strict'
CSRF_COOKIE_NAME = 'MyCookie'
CSRF_COOKIE_HTTPONLY = False
```

```
CSRF_COOKIE_SECURE = False
CSRF_TRUSTED_ORIGINS = []
```

编写一个新的视图用于测试 CSRF 配置:

```python
def test(request):
    from mysite import settings
    return render(request, 'polls/index.html', {})
```

在 polls/urls.py 中增加 URL:

```python
path('test/', views.test, name='test'),
```

修改 polls/templates/polls/index.html:

```
{% load i18n %}
{% load static %}
<link rel="stylesheet" type="text/css" href="{% static 'polls/style.css' %}">
<script src="https://apps.bdimg.com/libs/jquery/2.1.4/jquery.min.js"></script>
<script src="https://cdn.bootcss.com/jquery-cookie/1.4.1/jquery.cookie.min.js"></script>
<script>

$.ajaxSetup({
  data: {csrfmiddlewaretoken: '{{ csrf_token }}' },
});

function submit(){
    console.log("MyCookie:"+$.cookie("MyCookie"))
    $.ajax({
      url:"/polls/test/",
      type:"POST",
      data:{},
      headers:{
          "X-CSRF-AARONTOKEN":$.cookie("MyCookie")
       },
      success:function (arg) {
          console.log("success")
      },
      error:function(e){
        console.error(e)
      }
    })
}

</script>

{% if latest_question_list %}
    <ul>
    {% for question in latest_question_list %}
        <li><a href="{% url 'polls:detail' question.id %}">{{ question.question_text }}</a></li>
    {% endfor %}
    </ul>
{% else %}
    <!-- 还没有调查问卷! -->
```

```
      <p>{% trans "warningmsg" %}</p>
{% endif %}

<input type="button" id="btn1" onclick="submit()" value="测试 CSRF Cookie" />
```

启动网站，polls 页面 http://127.0.0.1:8000/polls/ 如下图所示。

打开浏览器的开发者工具（大部分浏览器可以通过按 F12 快捷键打开），以 Chrome 浏览器为例，CSRF Cookie 如下图所示。

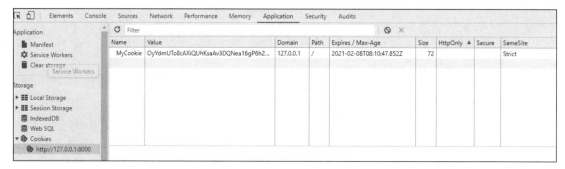

单击按钮"测试 CSRF Cookie"，查看开发者工具中的 Console 标签，如下图所示。

可以看到输出了 CSRF Cookie。

修改 CSRF_COOKIE_HTTPONLY = True，重复以上步骤，查看 Console 输出情况，如下图所示。

此时输出 undefined，说明前端 JavaScript 脚本已经没有权限访问 CSRF Cookie 了。

代码分析：

为 ajax 请求添加 CSRF Cookie，保证请求成功。

```
$.ajaxSetup({
    data: {csrfmiddlewaretoken: '{{ csrf_token }}' },
});
```

5.12 模型

5.12.1 ABSOLUTE_URL_OVERRIDES

默认值：{}

字典类型，Key 是 app_label.model_name 格式字符串，value 是一个方法，这个方法接收一个 Django 模型对象并返回对应的 URL。通过 ABSOLUTE_URL_OVERRIDES 可以为每一个 Django 程序添加或重写 get_absolute_url()。

代码示例：

```
ABSOLUTE_URL_OVERRIDES = {
    'blogs.weblog': lambda o: "/blogs/%s/" % o.slug,
    'news.story': lambda o: "/stories/%s/%s/" % (o.pub_year, o.slug),
}
```

不管对应的 model 是大写形式还是小写形式，Key 中的 model 字符串必须使用小写形式。

5.12.2 FIXTURE_DIRS

默认值：[]

一组保存 fixture 文件的目录，是对 fixtures 文件夹的补充。Django 按照目录在列表中出现的顺序进行查找。

列表中的目录需要使用 UNIX 风格书写，即使是在 Windows 系统中也要使用斜杠 "/"。

5.12.3 INSTALLED_APPS

默认值：[]

一组当前 Django 项目可用的应用程序，每一个列表项都是一个 Python 路径：一个应用程序的配置类（例如 polls 的配置类 polls.apps.PollsConfig）或者一个包含应用程序的包。

5.13 日志

5.13.1 LOGGING

默认值：一个日志配置字典

日志配置信息，LOGGING 的内容将会被当作参数传递给 LOGGING_CONFIG 所指向的方法。默认配置信息存储在 django/utils/log.py。

5.13.2 LOGGING_CONFIG

默认值：`'logging.config.dictConfig'`

该参数指向一个用于配置日志系统的可调用对象的路径。默认使用 Python 的 dictConfig：logging.config.dictConfig。

如果 LOGGING_CONFIG 设置为 None，系统将跳过日志配置过程。

5.14 模板

5.14.1 TEMPLATES

默认值：`[]`

一个包含所有模板引擎配置信息的数组，数组中的每一项都是一个包含模板引擎配置信息的字典。

下面是一个最简单的模板配置，通过这个配置，模板引擎会在所有已安装的应用程序的 templates 子目录中查找模板文件：

```
TEMPLATES = [
    {
        'BACKEND': 'django.template.backends.django.DjangoTemplates',
        'APP_DIRS': True,
    },
]
```

下面是 Polls 应用程序的模板配置：

```
TEMPLATES = [
    {
        'BACKEND': 'django.template.backends.django.DjangoTemplates',
        'DIRS': [],
        'APP_DIRS': True,
        'OPTIONS': {
            'context_processors': [
                'django.template.context_processors.debug',
                'django.template.context_processors.request',
                'django.contrib.auth.context_processors.auth',
                'django.contrib.messages.context_processors.messages',
            ],
        },
    },
]
```

参数介绍如下。

1. BACKEND

默认值：`Not defined`

模板引擎的 Python 路径。Django 内置的模板引擎包括以下两个：

```
'django.template.backends.django.DjangoTemplates'
'django.template.backends.jinja2.Jinja2'
```

> **注意**
> 模板引擎的 Key（本例中为 'BACKEND'）用于渲染模板时查找模板引擎，Key 必须唯一，即当存在多个模板引擎时只能有一个叫作 'BACKEND'。

2. DIRS

默认值：`[]`

模板文件的存放路径，列表中的位置就是模板引擎查找模板文件的顺序。

3. APP_DIRS

默认值：`False`

规定模板引擎是否在已安装的应用程序中查找模板文件。通过 django-admin startproject 命令安装的 Django 应用程序的 APP_DIRS 的默认值是 True。

4. OPTIONS

默认值：`{}`

额外传递给模板引擎的参数，不同的模板引擎包含不同的参数，如 DjangoTemplates 支持 autoescape、context_processors、file_charset 等参数；Jinja2 支持 context_processors。

除了以上配置信息外，Django 的模板配置还支持国际化配置等，如 TIME_FORMAT、TIME_ZONE 等，这与前面介绍的国际化配置一致，在此不再赘述。

5.15 URLs

5.15.1 ROOT_URLCONF

默认值：`Not defined`

URL 配置的全路径，如前面示例中的 ROOT_URLCONF = 'mysite.urls'。可以通过修改 HttpRequest 对象的 urlconf 属性进行重写。

5.15.2 APPEND_SLASH

默认值：`True`

当 APPEND_SLASH 被设置为 True 时，如果请求的 URL 与 URLConf 中的任何模式字符串都不匹配，且请求的 URL 没有以斜杠（/）结束，那么 Django 会自动在 URL 后追加一个斜杠（/）进行重定向。需要注意的是，如果 POST 请求末尾因为缺少斜杠而发生重定向，请求的数据有可能在重定向的过程中丢失。

> **注意**
> 只有在安装 CommonMiddleware 中间件的情况下，APPEND_SLASH 才会起作用。

5.15.3 PREPEND_WWW

默认值：False

优先使用 WWW 作为 URL 前缀。

要想使 PREPEND_WWW 生效，必须确保已经安装了 CommonMiddleware。

5.16 其他

5.16.1 DEFAULT_EXCEPTION_REPORTER_FILTER

默认值：'django.views.debug.SafeExceptionReporterFilter'

当 HttpRequest 对象是 None 时所使用的默认异常过滤器类。

5.16.2 MIDDLEWARE

默认值：{}

Django 项目中所使用的中间件。详细信息请查看第 12 章。

第 6 章 后台管理页面

前面介绍过如何使用 admin.site.register() 方法向 Admin 后台程序添加模型，包括问卷和问卷选项。不仅如此，Django 还可以对 Admin 后台进行更丰富的定制化操作。

虽然 Django 的后台管理页面很强大，但是，如果需要处理流程相关的任务最好还是自己开发视图，因为 Django 的后台管理功能主要提供了面向模型的接口，对于复杂的流程操作并不擅长。

登录 Django 后台管理页面的账号是通过 createsuperuser 命令创建的。默认情况下，能够登录管理后台的账号需要将 is_superuser 或者 is_staff 属性设置为 True。

6.1 ModelAdmin 属性

前面提到，如果要使管理后台能够编辑一个模型，那么需要提前注册这个模型。这个注册的功能就是通过 ModelAdmin 实现的。通常 ModelAdmin 保存在应用程序的 admin.py 文件中。

下面是一个最简单的 ModelAdmin：

```
from django.contrib import admin
from myproject.myapp.models import Author

class AuthorAdmin(admin.ModelAdmin):
    pass
admin.site.register(Author, AuthorAdmin)
```

前面代码中 ModelAdmin 类（AuthorAdmin）没有实现任何功能，此时将会应用默认的 admin 行为。如果只需要使用默认的 admin 行为，就像前面代码那样使用空的 ModelAdmin 类，那么可以将代码简化成如下：

```
from django.contrib import admin
from myproject.myapp.models import Author

admin.site.register(Author)
```

除了 register 方法外，还可以使用 register 装饰器达到同样的效果，使用 register 装饰器修改以上代码：

```
from django.contrib import admin
from .models import Author

@admin.register(Author)
class AuthorAdmin(admin.ModelAdmin):
    pass
```

使用 register 装饰器可以同时注册多个模块，还可以指定自定义管理后台，例如：

```
from django.contrib import admin
from .models import Author, Editor, Reader
from myproject.admin_site import custom_admin_site

@admin.register(Author, Reader, Editor, site=custom_admin_site)
class PersonAdmin(admin.ModelAdmin):
    pass
```

需要注意的是，如果 ModelAdmin 类的初始化方法引用了类自身，例如 super (PersonAdmin, self).__init__(*args, **kwargs)，此时不能使用 register 装饰器，但是，如果初始化代码没有引用类，例如 super().__init__(*args, **kwargs)，此时仍然可以使用 register 装饰器。

例如，按照以下方式修改在线投票系统中注册 Choice 类的方法：

```
@admin.register(Choice)
class ChoiceAdmin(admin.ModelAdmin):
    def __init__(self, *args, **kwargs):
        super(ChoiceAdmin, self).__init__(*args, **kwargs)
        pass
```

重新执行项目将会报错，如下图所示。

```
 9    @admin.register(Choice)
10    class ChoiceAdmin(admin.ModelAdmin):
11        def __init__(self, *args, **kwargs):
D 12          super(ChoiceAdmin, self).__init__(*args, **kwargs)

发生异常: NameError
name 'ChoiceAdmin' is not defined
  File "D:\Code\django3\mysite\polls\admin.py", line 12, in __init__
    super(ChoiceAdmin, self).__init__(*args, **kwargs)
  File "D:\Code\django3\mysite\polls\admin.py", line 10, in <module>
    class ChoiceAdmin(admin.ModelAdmin):
  File "D:\Code\django3\mysite\manage.py", line 17, in main
    execute_from_command_line(sys.argv)
  File "D:\Code\django3\mysite\manage.py", line 21, in <module>
    main()
```

但是，如果删除 super 方法中的类名引用，项目又能正常执行了，代码如下：

```
@admin.register(Choice)
class ChoiceAdmin(admin.ModelAdmin):
    def __init__(self, *args, **kwargs):
        super().__init__(*args, **kwargs)
        pass
```

ModelAdmin 类额外提供了一些可选属性，这些可选属性能够改变管理后台的显示行为，下面介绍这些可选项。

6.1.1　date_hierarchy

可以将模型中的 DateField 或者 DateTimeField 字段赋值给 date_hierarchy，此时列表就可以按照时间过滤数据了。例如，在没有设置 date_hierarchy 时，question 模型在管理后台的显示样式如下图所示。

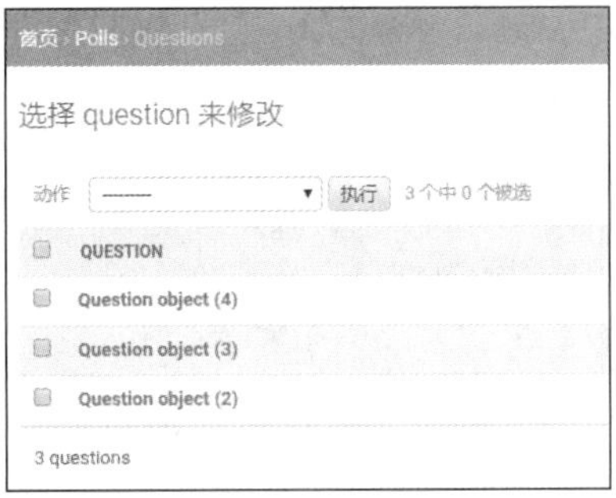

将问卷的发布时间字段赋值给 date_hierarchy，代码如下：

```
@admin.register(Question)
class QuestionAdmin(admin.ModelAdmin):
    date_hierarchy = 'pub_date'
    pass
```

此时管理后台多了发布时间的过滤功能，如下图所示。

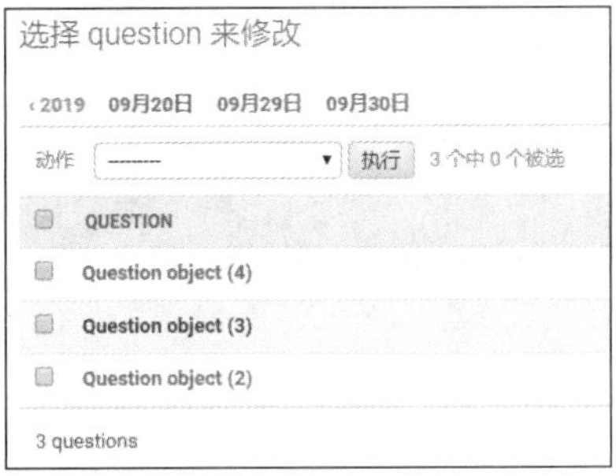

单击选择任意时间将会过滤出当天发布的问卷，如下图所示。

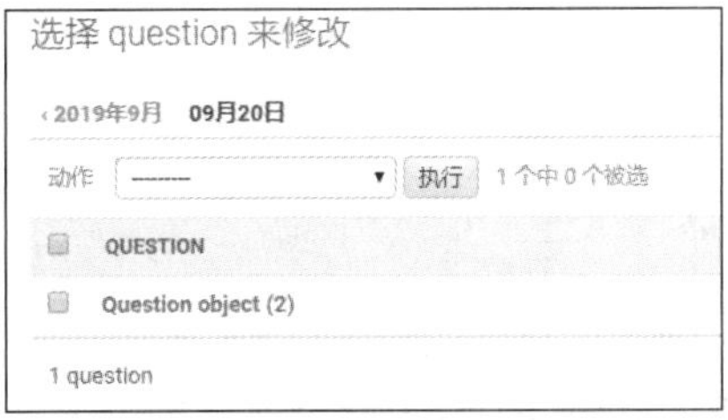

如果需要根据关联表的时间进行过滤，可以使用双下画线指定 date_hierarchy，如根据问卷的发布时间过滤问卷选项，可以按照如下方式指定 date_hierarchy：

```
@admin.register(Choice)
class ChoiceAdmin(admin.ModelAdmin):
    date_hierarchy = 'question__pub_date'
    def __init__(self, *args, **kwargs):
        super().__init__(*args, **kwargs)
        pass
```

需要注意的是，这里设置 date_hierarchy 时所使用的"question"是模型 Choice 定义时所声明的外键"question"，而不是"Question"模型。

此时问卷选项表"Chioces"的显示效果如下图所示。

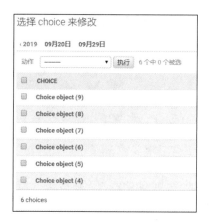

6.1.2 actions_on_top/actions_on_bottom

控制动作条的显示位置，默认在页面顶部显示动作条（如前面问卷管理中的"动作"下拉列表）。如需修改动作条的显示位置，只要在 ModelAdmin 进行设置即可，如调整动作条在页面底部显示：

```
@admin.register(Question)
class QuestionAdmin(admin.ModelAdmin):
    date_hierarchy = 'pub_date'
    actions_on_top = False
    actions_on_bottom = True
    pass
```

显示效果如下图所示。

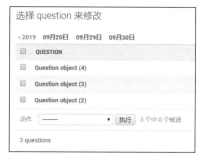

如果同时将 actions_on_top 和 actions_on_bottom 设置为 False，则不显示动作条。

6.1.3 actions_selection_counter

控制动作条中是否显示计数功能，默认显示（如前面问卷管理中的"动作"条后显示被选中数据量）。如果需要取消显示计数功能，可以在 ModelAdmin 中将 actions_selection_counter 设置为 False：

```
@admin.register(Question)
class QuestionAdmin(admin.ModelAdmin):
    date_hierarchy = 'pub_date'
    actions_on_top = False
    actions_on_bottom = True
    actions_selection_counter = False
    pass
```

显示效果如下图所示。

6.1.4 empty_value_display

空数据的显示方式。如果一条记录的某个字段值是 None 或者空字符串，将使用 empty_value_display 代替该字段值，默认值为"-"。

为演示方便，修改 Question 模型，增加 description 字段：

```
class Question(models.Model):
    question_text = models.CharField(max_length=200)
    pub_date = models.DateTimeField('date published')
    description = models.CharField(max_length=200, null=True, blank=True)
```

执行 Migrations 命令使变更生效，如果数据库迁移过程中出现错误，可以先使用 migrate zero 来清空全部数据结构，然后再重新迁移数据库：

```
python manage.py migrate polls zero
```

> **注意**
>
> migrate zero 会将应用程序对应的全部数据结构清空，在生产环境中一定要谨慎使用该命令。

修改完数据库结构后，可以看一下 empty_value_display 的使用效果。首先在 ModelAdmin 类中增加 list_display 属性，关于这个属性的用法在后续内容中详细介绍。此时 QuestionAdmin 类如下：

```
@admin.register(Question)
class QuestionAdmin(admin.ModelAdmin):
    date_hierarchy = 'pub_date'
    actions_on_top = False
    actions_on_bottom = True
    actions_selection_counter = False

    list_display = ('question_text', 'pub_date', 'description')
    pass
```

重启网站，此时在管理后台查看问卷列表，可以看到新的问卷列表已经不再显示"Question Object"，而是显示问卷的真实信息，而 DESCRIPTION 字段显示的是一个横线（"-"），如下图所示。

这个横线就是 empty_value_display 所指定的值，如果修改 empty_value_display 会怎样呢？按照以下方式修改代码：

```
@admin.register(Question)
class QuestionAdmin(admin.ModelAdmin):
    ...
    empty_value_display = 'NULL'
    ...
```

刷新运营后台页面，可以看到 DESCRIPTION 已经显示自定义的空值，如下图所示。

6.1.5 exclude

该字段指定一组从表单中移除的字段。例如，创建新的调查问卷时，通常需要使用当前系统时间作为发布时间，将 pub_date 从表单中移除：

```
@admin.register(Question)
class QuestionAdmin(admin.ModelAdmin):
    ...
    exclude = ('pub_date',)
    ...
```

从表单中移除 pub_date，换句话说，就是在表单中保留其他字段，以下代码的效果与上述代码的相同。

```
@admin.register(Question)
class QuestionAdmin(admin.ModelAdmin):
    ...
    fields = ('question_text', 'description',)
    ...
```

此时刷新管理后台，添加问卷的页面显示效果如下图所示。

6.1.6 fields

fields 属性除了可以控制显示的字段外，还可以控制字段的显示顺序，以及对多个字段进行分组。例如，可以通过改变字段在元组中的顺序来调整字段在页面上的显示顺序：

```
@admin.register(Question)
class QuestionAdmin(admin.ModelAdmin):
    ...
    fields = ('pub_date', 'question_text', 'description')
    ...
```

显示效果如下图所示。

有时需要将意义相近的一组属性显示在一行，可以使用嵌套元组的方式设置。例如，将 'question_text' 和 'description' 显示在一行可以按照以下方式编写代码：

```
@admin.register(Question)
class QuestionAdmin(admin.ModelAdmin):
    ...
    fields = ('pub_date', ('question_text', 'description'))
    ...
```

显示效果如下图所示。

6.1.7 fieldsets

fieldsets 也可以用来控制字段在管理页面的显示布局，fieldsets 是一个元组，这个元组的每一项都是一个拥有两个元素的元组，每一个二级元组都对应管理页面上的一个片段。二级元组的格式如下：

```
(name, field_options)
```

这里的 name 是每一个片段的显示名字，field_options 是一组需要显示的字段，包括字段的显示行为。下面使用 fieldsets 重新编写 QuestionAdmin 类：

```
@admin.register(Question)
class QuestionAdmin(admin.ModelAdmin):
    ...
    fieldsets = (
        ('必填项', {
            'fields': ('question_text', 'description',),
        }),
        ('自动生成项', {
            'classes': ('collapse',),
            'fields': ('pub_date',)
        })
    )
    ...
```

刷新管理页，效果如下图所示。

可以看到所有字段被分成了两组显示：其中 'question_text' 和 'description' 显示在"必填项"中，而 'pub_date' 显示在"自动生成项"中。"自动生成项"是一个可折叠区域，单击"显示"可以展开内容，如下图所示。

单击"隐藏"又可以隐藏该区域。

不要同时指定 fieldsets 和 fields，否则会出现如下图所示的异常。

```
ERRORS:
<class 'polls.admin.QuestionAdmin'>: (admin.E005) Both 'fieldsets' and 'fields' are specified.
```

field_options 字典可以包含以下 Key。

fields

这里的 fields 与 ModelAdmin 类的 fields 属性用法基本一致，用于控制显示的字段以及对字段进行分组。

classes

一组 CSS 类，这些类将会被应用于 fieldset。

默认的管理后台提供了很多 CSS 类，其中最常用的两个是 collapse 和 wide。collapse 类允许 fieldset 被折叠和展开。wide 类可以为 fieldset 提供额外的水平空间，效果如下图所示。

description

为每一个 fieldset 提供的额外的描述信息。需要注意的是，description 不仅仅支持普通字符串，还可以接收 HTML 片段，修改前面的 fieldsets 代码：

```
fieldsets = (
    ('必填项', {
        'fields': ('question_text', 'description',),
    }),
    ('自动生成项', {
        'classes': ('collapse',),
        'description':'<span style="color:red">这里的所有字段都不需要赋值</span>',
        'fields': ('pub_date',)
    })
)
```

刷新管理页面，效果如下图所示。

6.1.8 filter_horizontal

用于优化多对多关系在管理后台的显示方式。为方便演示，这里增加一个问卷类别模型，每个问卷类别都可以包含多个问卷，代码如下：

```
class Category(models.Model):
    order_number = models.CharField(max_length=20)
    question_name = models.ManyToManyField(Question)
```

接下来在 admin.py 文件中添加对应的 ModelAdmin 类：

```
from .models import Category

@admin.register(Category)
class CategoryAdmin(admin.ModelAdmin):
    pass
```

重启管理后台，新增一个问卷分类，效果如下图所示。

在此可以看到创建分类时可以指定多个问卷，问题貌似已经得到很好解决，但是想象一下，如果当前系统已经运行很久，很多问卷需要归类，那么仍然从下拉列表中选择就会非常困难，此时可以使用 filter_horizontal 进行改善。

修改 ModelAdmin：

```
@admin.register(Category)
class CategoryAdmin(admin.ModelAdmin):
    filter_horizontal = ('question_name',)
    pass
```

需要注意的是，filter_horizontal 可接收的值必须是 Category 模型所包含的属性（因为在过滤时需要根据这些字段来筛选问卷），刷新管理后台，效果如下图所示。

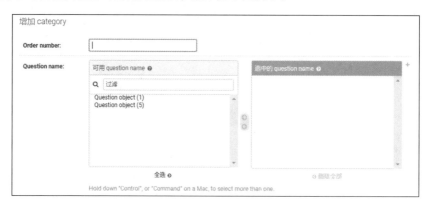

此时选择问卷就已经非常方便，但是问卷在列表中显示为"Question object"仍然是一个非常影响用户使用的问题，可以通过重写模型的 __str__ 方法的方式来解决，如为问卷类增加以下代码：

```
class Question(models.Model):
    ...
    def __str__(self):
        return self.question_text
```

重启网站，此时列表中已经能够正常显示问卷了，效果如下图所示。

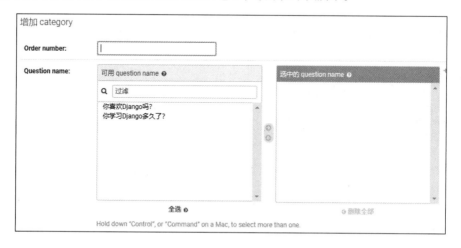

6.1.9 filter_vertical

filter_vertical 与 filter_horizontal 功能完全一样，只是在显示选择框时使用竖向排版，效果如下图所示。

6.1.10 form

Django 会为每一个模型动态创建一个 ModelForm，这个表单将会被用于模型的添加和编辑页面，前面提到的添加问卷页面就是一个 ModelForm。在 Django 中，你可以很轻松地编写自定义 ModelForm 来替代默认的 ModelForm，不过使用 ModelAdmin.get_form() 方法定制默认的 ModelForm 更方便。

下面使用 ModelForm 重写 QuestionAdmin：

```python
from django import forms
from django.contrib import admin
from .models import Question, Choice, Category

class QuestionForm(forms.ModelForm):
    class Meta:
        model = Question
        exclude = [' description']

@admin.register(Question)
class QuestionAdmin(admin.ModelAdmin):
    exclude = ['pub_date']
    form = QuestionForm
    pass
```

刷新管理后台，此时添加问卷的页面显示效果如下图所示。

如果在 ModelForm 中定义了 Meta.model，那么必须同时定义 Meta.exclude 或者 Meta.fields。需要注意的是，如果 ModelForm 和 ModelAdmin 同时定义了 exclude 或者 fields，那么 ModelAdmin 中的 exclude 或者 fields 优先级更高，一般不必在 ModelForm 中定义 Meta.model。

6.1.11 formfield_overrides

formfield_overrides 可以作为应急方式为管理后台重写 field 属性。formfield_overrides 的值是一个字典，字典的键是模型的字段类，例如 models.TextField，字典的值仍然是一个字典，这个字典用于重写模型的字段的属性，新的属性会在构造模型字段时使用。

下面使用 formfield_overrides 属性为问卷类的每个文本字段增加帮助字符串：

```python
from django.db import models

@admin.register(Question)
class QuestionAdmin(admin.ModelAdmin):
    ...
    formfield_overrides = {
```

```
            models.CharField: {'help_text': '<span style="color:red;">请输入</span>'},
        }
        pass
```

刷新管理页面，可以看到在文本框下已经有了帮助文本，如下图所示。

对于内嵌字典可以接收哪些键，可以参考模型字段的构造函数。formfield_overrides 的键是真正的模型字段类，千万不要使用字符串。最后需要注意的是，如果模型的外键（如 ForeignKey 或者 ManyToManyField）已经被放置在 raw_id_fields、radio_fields 或者 autocomplete_fields 中，那么就不能修改这些字段的显示控件了。

6.1.12　inlines

ModelAdmin.inlines 属性配置用户在添加、编辑模型时可以关联的对象，例如在编辑调查问卷的同时可以编辑问卷选项。

详细说明请参考 6.5 节的 InlineModelAdmin 对象和 6.2.24 节的 ModelAdmin.get_formsets_with_inlines() 方法。

6.1.13　list_display

对于这个属性，在 6.1.4 节中已经介绍过，list_display 可以控制模型列表中显示的字段。

如果没有设置 list_display，那么 Django 管理后台会使用模型对象的 __str__() 作为唯一列显示，默认的 __str__() 如下：

```
def __str__(self):
    return '%s object (%s)' % (self.__class__.__name__, self.pk)
```

list_display 可以接收 4 种类型的值：
❑ 模型字段名，例如：

```
@admin.register(Question)
class QuestionAdmin(admin.ModelAdmin):
    list_display = ('question_text', 'pub_date', 'description')
```

❑ 可调用对象，该对象的唯一参数是模型实例，例如：

```
def star_case_name(obj):
    return (" * %s" % obj.question_text)
star_case_name.short_description = '标题'
```

```python
@admin.register(Question)
class QuestionAdmin(admin.ModelAdmin):
    list_display = (star_case_name,'description',)
```

显示效果如下图所示。

- 一个 ModelAdmin 方法的名字，这个方法唯一的参数是模型实例，例如：

```python
@admin.register(Question)
class QuestionAdmin(admin.ModelAdmin):
    list_display = ('star_case_name','description',)

    def star_case_name(self, obj):
        return (" * %s" % obj.question_text)
    star_case_name.short_description = '标题'
```

显示效果与可调对象的相同。

- 一个字符串，该字符串可以是模型的属性或者方法，如果是方法则不能接收任何参数。

```python
class Question(models.Model):
    ...
    def guid(self):
        import uuid
        return uuid.uuid4()
    guid.short_description = 'guid'


@admin.register(Question)
class QuestionAdmin(admin.ModelAdmin):
    ...
    list_display = ('guid', 'question_text', 'pub_date', 'description',)
```

显示效果如下图所示。

使用 list_display 时需要注意以下几点。
- 如果待显示字段是一个外键（ForeignKey），那么 Django 显示 __str__() 方法值。
- 不支持多对多字段（ManyToManyField），如果一定要显示多对多字段，最好编写一个自定义方法，然后在 list_display 中使用自定义方法名。
- 对于布尔字段（BooleanField），Django 显示 on / off 图标。
- 如果给出的字符串是模型或者 ModelAdmin 的方法或者其他可调用对象，返回的字符串包含 HTML 标记，Django 会原样显示这些标记，如果需要按照 HTML 样式显示这些标记，请使用 format_html()，修改 Question 模型：

```
from django.utils.html import format_html

class Question(models.Model):
    question_text = models.CharField(max_length=200)
    pub_date = models.DateTimeField('date published')
    description = models.CharField(max_length=200, null=True, blank=True)
    def color_name(self):
        return format_html('<span style="color: #f00;">' + self.question_text + '</span>')

    def color_description(self):
        if self.description:
            return '<span style="color: #f00;">' + self.description + '</span>'
        else:
            return '<span style="color: #f00;">没有任何描述信息</span>'
```

修改 QuestionAdmin：

```
list_display = ('color_name','color_description',)
```

刷新管理页面，效果如下图所示。

- 对于可调用方法（包括模型或者 ModelAdmin 的方法），可以使用 short_description 修改列标题。
- 当字段值是 None、空字符串或者空列表时，Django 显示 " - "，可以使用 AdminSite.empty_value_display 修改默认的显示值。
- 如果模型方法、ModelAdmin 或者其他可调用方法返回 True 或者 False，并且该方法的 boolean 属性是 True 时，Django 显示对应的 on/off 图标。
- 如果方法返回布尔值，Django 显示 on / off 图标。
- __str__() 方法与其他模型方法一样，在 list_display 中也可以使用 __str__，如 list_display = ('__str__', 'some_other_field')。
- 如果 list_display 中的值不是真正的数据库字段，这些字段不能用于排序，但是可以使用方法的 admin_order_field 属性进行排序：

```
from django.contrib import admin
class Question(models.Model):
    ...
    def color_name(self):
        return format_html('<span style="color: #f00;">' + self.question_text + '</span>')

    color_name.admin_order_field = 'question_text'
```

如果需要进行倒序排列，可以在字段名前添加一个横线：

`color_name.admin_order_field = '-question_text'`

如果需要根据关联表的字段进行排序，可以使用双下画线，例如：

`color_name.admin_order_field = '关联表的属性名__关联表字段'`

admin_order_field 还支持简单的查询表达式：

`color_name..admin_order_field = Concat('字段1', '字段2')`

- list_display 中的字段名同时也是字段对应的 CSS 类，例如字段 color_description 将会有 CSS 类 column-color_description。这可以方便帮助开发人员针对每一列进行样式设置，如下图所示。

```
▼<thead>
  ▼<tr>
    ▶<th scope="col" class="action-checkbox-column">…</th>
    ▶<th scope="col" class="sortable column-color_name">…</th>
    ▶<th scope="col" class="column-color_description">…</th>
   </tr>
 </thead>
```

- Django 解释 list_display 内容的顺序如下：
- 模型字段。
- 可调用方法。
- ModelAdmin 属性名。
- 模型属性名。

> **注意**
>
> Django 官网多次提到 model field、model attribute 和 ModelAdmin attribute，其中 field 表示数据库中的一列，而 attribute 表示的是 model 的方法或 property。property 可以通过 property() 方法来声明。

6.1.14 list_display_links

默认情况下，list_display 中的第一项会被显示为链接形式，在列表页单击该链接可以进入编辑页面，如前面通过单击每一个问卷的标题列可以进入编辑页面。

list_display_links 可以改变以上行为：

- 将 list_display_links 设置为 None 则不显示任何链接。
- 以元组或者列表的形式指定一组字段（与 list_display 一致），这些字段都将显示为链接。

Django 允许定义任意多个字段作为链接，唯一的要求就是 list_display_links 中的内容必须提前在 list_display 中定义好。

如需要将 color_name 和 color_description 作为编辑页面的入口，可以按照以下方式编写代码：

```
list_display = ('color_name','color_description',)
list_display_links = ('color_name','color_description',)
```

6.1.15　list_editable

通过设置 list_editable 可以在模型列表页直接编辑字段。

使用 list_editable 的时候需要注意以下几点。

❑ list_editable 中的字段必须同时出现在 list_display 中。

❑ 同一个字段不能同时出现在 list_editable 和 list_display_links 中。

修改 Choice 模型类和 ChoiceAdmin 类如下：

```
class Choice(models.Model):
    question = models.ForeignKey(Question, on_delete=models.CASCADE)
    choice_text = models.CharField(max_length=200)
    votes = models.IntegerField(default=0)

    def question_text(self):
        return self.question.question_text

@admin.register(Choice)
class ChoiceAdmin(admin.ModelAdmin):
    date_hierarchy = 'question__pub_date'
    list_display = ('question_text','choice_text','votes')
    list_display_links = ('question_text',)
    list_editable = ('choice_text',)
    def __init__(self, *args, **kwargs):
        super().__init__(*args, **kwargs)
        pass
```

重启 Django 应用，在后台查看 choice 列表，如下图所示。

可以看到此时 CHOICE TEXT 列处于可编辑状态，同时修改 3 条记录保存，如下图所示。

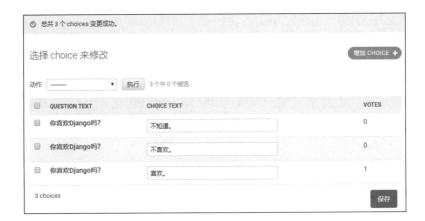

6.1.16 list_filter

通过设置 list_filter，可以在管理后台的右边栏激活过滤器插件。

例如为 QuestionAdmin 增加以下代码：

```
list_filter = ('question_text',)
```

此时刷新后台管理页面将会看到在页面右侧显示了过滤器，如下图所示。

设置 list_filter 时需要注意以下几点：
- 如果待设置的元素是模型字段，只能是以下类型：BooleanField、CharField、DateField、DateTimeField、IntegerField、ForeignKey 和 ManyToManyField。
- list_filter 中的字段也可以使用双下画线，例如：question__question_text。
- 如果 list_filter 所使用的元素是一个继承自 django.contrib.admin.SimpleListFilter 的类，这个类必须给出 title 和 parameter_name，同时要重写 lookups 和 queryset 方法。
- 可以通过元组对被搜索元素进行限制，如：

```
@admin.register(Choice)
class ChoiceAdmin(admin.ModelAdmin):
    ...
    list_filter = (
        ('choice_text', admin.AllValuesFieldListFilter),
)
```

6.1.17 list_per_page

每一个模型的列表页所能显示的最大项目数，默认每页最多显示 100 条记录。

例如为 ChoiceAdmin 设置 list_per_page = 2，列表页将会显示分页按钮，如下图所示。

6.1.18 list_max_show_all

显示全部记录的阈值。当模型现有的记录总数小于或等于 list_max_show_all 时，列表页将会显示一个 Show all 链接，单击链接显示全部记录。list_max_show_all 的默认值是 200。

继续修改 ChoiceAdmin，添加代码：list_max_show_all = 10，刷新网站，可以看到"显示全部"按钮，如下图所示。

单击"显示全部"按钮将会显示全部问卷选项。

6.1.19 list_select_related

list_select_related 属性可以告诉 Django 在提取模型列表时是否使用 select_related() 方法。select_related() 方法使得 Django 能够创建一个复杂的查询语句，该查询可以同时查找模型关联的外键对象，减少访问数据库的次数，这在一定程度上提升了 SQL 性能。

list_select_related 可以接收布尔值或者一组外键，默认值是 False。当 list_select_related 设置为 True 时，Django 将会一直使用 select_related() 方法，当 list_select_related 设置为 False 时，Django 将会查看 list_display，如果 list_display 包含任何外键则调用 select_related() 方法。

使用元组或列表是一个更友好的方式，空元组或列表能够阻止 Django 调用 select_related() 方法，否则元组或者列表值将会被传递给 select_related() 方法。

如果希望根据不同的请求动态指定 list_select_related，那么可以通过调用 get_list_select_related() 方法来实现。

如果列表页的 QuerySet 已经调用了 select_related()，那么 ModelAdmin 将会忽略 list_select_related 属性。

6.1.20 ordering

用于对管理后台的模型列表进行排序显示，可以接收一个列表或元组。如果没有提供 ordering，Django 会使用模型的默认排序。

可以通过 get_ordering() 方法来实现动态排序。

6.1.21 paginator

用于分页的类对象，默认值为 django.core.paginator.Paginator。如果使用自定义分页类，同时自定义分页类与默认分页类结构不一致，还需要实现 ModelAdmin.get_paginator() 方法。

6.1.22 prepopulated_fields

通过设置 prepopulated_fields 属性可以使得用户在管理后台输入某些字段时自动填充另一个字段。例如，使用下面方法可以使得用户在输入调查问卷标题时自动填充描述信息：

```
@admin.register(Question)
class QuestionAdmin(admin.ModelAdmin):
    ...
    prepopulated_fields = {"description": ("question_text",)}
    ...
```

网页显示效果如下图所示。

当用户在 Question text 文本框输入 this is a good book 时，Description 字段自动填充了 good-book。具体实现原理是，当用户在源字段输入文字时，Django 前端代码调用一段 JavaScript 脚本，该脚本将原始字符串中的空格替换成短横线，将原始字符串中的大写字母转换成小写字母，将各种英语停顿词（如 'a'、'an'、'as'）删除。

> **注意**
>
> 以上行为都是在用户输入时完成的，不是用户保存数据后才生成的，同时 prepopulated_fields 属性不能接收 DateTimeField、ForeignKey、OneToOneField 和 ManyToManyField 这几种类型的字段。

最后，由于 JavaScript 脚本只是用来调整英文字符串，因此，如果输入内容包含中文将不会生成目标字符串。

6.1.23 preserve_filters

当新增、编辑、删除结束回到列表页时保留原来的搜索条件。可以通过将 preserve_filters 设置为 False 来取消保存搜索条件。

如下图所示，当启用搜索条件"你喜欢 Django 吗 1？"时，每次新增问卷结束回到列表页时，列表内容仍然是使用搜索条件搜索出来的数据。

6.1.24 radio_fields

通过设置 radio_fields 可以在新增或者编辑页面使用单选按钮显示外键或者 choices 集合（默认情况下使用下拉列表）。

下面代码是使用单选按钮显示问卷选项的外键：

```
@admin.register(Choice)
class ChoiceAdmin(admin.ModelAdmin):
    ...
    radio_fields = {'question':admin.VERTICAL}
    ...
```

显示效果如下图所示。

通过 admin.HORIZONTAL 可以横向展示全部外键。

6.1.25 autocomplete_fields

autocomplete_fields 是一组外键或者多对多字段（ForeignKey、ManyToManyField），使用自动完成功能可以快速地在多个数据中定位到想要的值。

考虑以下场景，当调查问卷系统已经运行了很久，系统中存在很多问卷，此时需要对某一个问卷新增选项，在原来的系统中使用下拉列表来查找对应的调查问卷，这将会非常困难，而使用 autocomplete_fields 可以通过输入关键字的方式快速找到准确的问卷。

Django 通过 Select2.js 完成以上功能，外观上 autocomplete_fields 与原来的下拉列表非常相似，不过多了一个输入框,当用户输入内容时,Select2.js 向后台服务发送一个异步请求,从服务器端提取相应内容。

使用 autocomplete_fields 时必须在对应外键的 ModelAdmin 类中定义 search_fields,Select2.js 将会使用 search_fields 搜索数据。

为了防止数据泄漏,用户必须有对相关对象的查看和修改权限。

下面修改代码,增加对问卷选项的自动填充功能:

```
@admin.register(Question)
class QuestionAdmin(admin.ModelAdmin):
    ...
    search_fields = ('question_text',)
    ...

@admin.register(Choice)
class ChoiceAdmin(admin.ModelAdmin):
    ...
    autocomplete_fields = ('question',)
    ...
```

重启网站服务,新增问卷选项,可以看到 Question 字段已经有了自动填充功能,如下图所示。

6.1.26 raw_id_fields

当用户对外键比较了解时,很多用户更愿意自己输入外键对应的值而不是在下拉列表中选择。例如,当调查问卷系统中只有两个问卷并且用户对这两个问卷非常熟悉,那么在新增问卷选项的时候用户更愿意直接输入问卷 id。另外在下拉列表中显示过多的外键也可能导致性能问题,所以很多时候使用文本框代替下拉框。

```
@admin.register(Choice)
class ChoiceAdmin(admin.ModelAdmin):
    ...
    raw_id_fields = ('question',)
    ...
```

效果如下图所示。

当外键是 ForeignKey 时，用户只能输入对应的主键，而当外键是 ManyToManyField 时，用户需要输入以英文逗号分隔的一组主键，如下图所示。

6.1.27　readonly_fields

默认情况下，管理后台会将模型的全部字段以可编辑的形式显示。任何包含在 readonly_fields 中的字段都会原样显示并且不可编辑，同时这些字段会排除在 ModelForm 之外（编辑时不会随着表单被提交到服务器）。

readonly_fields 除了可以显示静态数据外，还可以显示模型方法或者 ModelAdmin 的方法，这与 list_display 一样。

6.1.28　save_as

使用 save_as 可以使得在编辑一个现有条目时，将编辑内容保存为一个新条目，例如前面编辑问卷选项时页面结构如下图所示。

在 ChoiceAdmin 中添加代码：save_as = True。重新编辑问卷选项，可以看到原来"保存并增加另一个"按钮变成"保存为新的"，效果如下图所示。

6.1.29　save_as_continue

当 save_as = True 时，单击"保存为新的"按钮后默认会停留在编辑页，如果将 save_as_continue 设置为 False，单击"保存为新的"按钮后页面将会返回到列表页。

save_as_continue 的默认值是 True。

6.1.30 save_on_top

设置 save_on_top=True 后，保存按钮将同时显示在页面顶端和页面底端，效果如下图所示。

6.1.31 search_fields

使用 search_fields 属性为列表添加可搜索字段，列表页显示搜索框。可以用于 search_fields 属性的字段应该是文本类型字段，如 CharField 或者 TextField。在 search_fields 中也可以通过双下画线的方式查找外键（ForeignKey 或者 ManyToManyField）关联的字段值：

```
search_fields = ['foreign_Key__related_fieldname']
```

Django 在执行搜索的过程中会将用户输入的搜索条件拆分成一组单词，如果每一个单词都存在于一个搜索字段中，那么搜索结果为真。搜索过程中不区分字母大小写（icontains 查询）。

例如，根据用户姓、名进行搜索，search_fields = ['first_name', 'last_name']，搜索条件为 john lennon，那么 Django 后台执行的 SQL 语句如下：

```
WHERE (first_name ILIKE '%john%' OR last_name ILIKE '%john%')
AND (first_name ILIKE '%lennon%' OR last_name ILIKE '%lennon%')
```

如果不希望使用默认的 icontains 查询，也可以在指定查询字段时给出查询类型，例如针对名使用精确查找，可以在 first_name 字段后添加 exact：

```
search_fields = ['first_name__exact']
```

在此需要注意的是，因为 Django 在执行查询时会将搜索条件拆分成一组单词，然后使用 AND 将每个查询条件关联起来，所以，如果使用 exact 查询，只能给出一个查询字段，否则查询结果可能不正确。例如，john 存在于 first_name，而 lennon 存在于 last_name 中，那么执行以下语句都为真：

```
first_name = '%john%' OR last_name ILIKE '%john%'
first_name = '%lennon%' OR last_name ILIKE '%lennon%'
```

但期望结果是 john 和 lennon 两个单词同时存在于 first_name 中，因此，以上查询结果不准确。
search_fields 支持下表所示的查询类型，列表中的前缀与后缀功能一样。

前缀	后缀
^	startswith
=	iexact
@	search
空（None）	icontains

修改 QuestionAdmin，添加搜索字段：

```
@admin.register(Question)
class QuestionAdmin(admin.ModelAdmin):
    ...
    search_fields = ('^question_text',)
    ...
```

重启网站，此时页面顶部显示搜索框，如下图所示。

分别输入"你"和"Django"查看搜索结果，如下图所示。

由此可见，Django 能够正确使用 startswith 搜索数据。

如果 Django 提供的默认搜索方式不能满足产品需求，可以通过编写 ModelAdmin.get_search_results() 方法来实现定制化搜索。

6.1.32 show_full_result_count

表示是否显示搜索结果的总数，如前面搜索"你"的时候显示"2 条结果。(总共 3)"。如果 show_full_result_count 设置为 False，同样搜索条件将会显示"2 条结果。(显示全部)"，如下图所示。

6.1.33 sortable_by

默认情况下，list_display 中的所有字段在列表页中都可以排序。当指定 sortable_by 时，只有 sortable_by 中的字段可以排序，sortable_by 是 list_display 的子集。当 sortable_by 为空集合时，全部字段禁用排序功能。

可以使用 get_sortable_by() 方法自定义排序规则。

```
@admin.register(Question)
class QuestionAdmin(admin.ModelAdmin):
    list_display = ('question_text','description','pub_date')
    sortable_by = ('question_text',)
```

6.1.34 view_on_site

控制是否在编辑页面显示"在站点上查看"按钮，单击该按钮可以跳转至对象页面。view_on_site 可以接收一个布尔值或者可执行对象（callable）。

当 view_on_site=True 时，模型的 get_absolute_url() 方法可以生成当前数据的 URL。

为 QuestionAdmin 增加 view_on_site = True，修改 Question 模型类，增加以下代码：

```
def get_absolute_url(self):
    return""# 仅用于演示 view_on_site 功能，因此没有给出具体的 URL。
```

重启网站，编辑任意调查问卷，可以看到网页上多了一个按钮（"在站点上查看"），如下图所示。

由于 get_absolute_url() 方法返回的是空字符串，因此，单击"在站点上查看"按钮网页将导航到首页（http://127.0.0.1:8000/）。

6.1.35 自定义模板

Django 允许开发人员使用自己的模板替代默认的管理后台模板，不同页面对应的模板也不一样，具体如下：

```
ModelAdmin.add_form_template
ModelAdmin.change_form_template
ModelAdmin.change_list_template
ModelAdmin.delete_confirmation_template
ModelAdmin.delete_selected_confirmation_template
ModelAdmin.object_history_template
ModelAdmin.popup_response_template
```

下面演示如何使用自定义模板。

首先添加自定义模板：

```html
# templates\polls\question_list.html
<style type="text/css">
table {border-collapse:collapse; }
table,th, td {border: 1px solid black;}
th {background-color:green; color:white;}
td {text-align:right;}
</style>
<table>
    <thead>
        <tr>
            <th> 标题 </th>
            <th> 描述 </th>
        <tr>
    </thead>

    <tbody>
    {% for question in other_objects%}
        <tr>
            <td>{{ question.question_text }}</td>
            <td>{{ question.description }}</td>
        <tr>
    {% endfor %}
    </tbody>
</table>
```

然后修改 QuestionAdmin：

```python
import os

@admin.register(Question)
class QuestionAdmin(admin.ModelAdmin):
    ...
    path = os.path.normpath(os.path.join(os.path.dirname(__file__),
        r"templates/polls/question_list.html"))
```

```
        change_list_template = path

    def changelist_view(self, request, extra_context={}):
        other_objects = Question.objects.all()
        extra_context['other_objects'] = other_objects
        return super().changelist_view(request, extra_context)
    ...
```

最后重启管理后台，在首页单击 Questions 进入问卷列表页，此时已经使用自定义模板了，效果如下图所示。

6.2 ModelAdmin 方法

6.2.1 save_model

方法签名：ModelAdmin.save_model(request, obj, form, change)

重写 save_model 方法可以使得用户在创建或修改对象时增加额外操作，其中，参数 request 是 HttpRequest 实例，obj 是模型实例，form 是 ModelForm 实例，change 是一个布尔值用于标记新增还是修改。

下面代码是在添加或者编辑调查问卷时判断用户是否输入了问卷描述信息，如果没有输入，则保存默认信息：

```
@admin.register(Question)
class QuestionAdmin(admin.ModelAdmin):
    ...
    def save_model(self, request, obj, form, change):
        if not obj.description:
            obj.description = "新增或编辑调查问卷"
        super().save_model(request, obj, form, change)
    ...
```

重启网站，添加一个新问卷，效果如下图所示。

单击"保存"按钮，回到列表页查看新添加的调查问卷，效果如下图所示。

6.2.2 delete_model

方法签名：ModelAdmin.delete_model(request, obj)

重写 delete_model 方法使得用户在删除对象时增加额外操作，使用方法与 save_model 相似，需要调用 super().delete_model() 完成删除操作。

6.2.3 delete_queryset

方法签名：ModelAdmin.delete_queryset(request, queryset)

delete_queryset 方法可用于删除指定的查询结果集中的全部对象。

6.2.4 save_formset

方法签名：ModelAdmin.save_formset(request, form, formset, change)

save_formset() 接收一个 HttpRequest 实例、父级 ModelForm 实例以及一个用于标记添加或者修改父级对象的布尔值。

使用 save_formset() 方法修改 QuestionAdmin 类，使得用户在编辑调查问卷的同时设置问卷选项的默认投票数为 0：

```
class ChoiceInline(admin.StackedInline):
    model = Choice

@admin.register(Question)
class QuestionAdmin(admin.ModelAdmin):
    ...
    inlines = (ChoiceInline,)
    def save_formset(self, request, form, formset, change):
        instances = formset.save(commit=False)
        for obj in formset.deleted_objects:
            obj.delete()
        for instance in instances:
            import datetime
            instance.votes = 0
            instance.save()
        formset.save_m2m()
```

6.2.5　get_ordering

方法签名：ModelAdmin.get_ordering(request)

get_ordering() 方法返回一个列表或者元组，与 ordering 属性的用法相似。

以下代码使得问卷列表能够根据登录用户是否是超级管理员进行不同的排序（该方法没有实际意义，仅作为演示使用）：

```python
@admin.register(Question)
class QuestionAdmin(admin.ModelAdmin):
    ...
    def get_ordering(self, request):
        if request.user.is_superuser:
            return ['question_text', 'description']
        else:
            return ['question_text']
    ...
```

6.2.6　get_search_results()

方法签名：ModelAdmin.get_search_results(request, queryset, search_term)

通过修改 get_search_results() 方法可以定制搜索结果。其中，queryset 是通过当前搜索条件搜索出来的结果，search_term 是用户提交的搜索条件。get_search_results() 返回一个包含两个元素的元组，第一个元素是新的搜索结果，第二个元素用于标志当前结果是否包含重复值。

下面修改 get_search_results() 方法使得当搜索条件存在空格时 Django 不拆分搜索条件：

```python
@admin.register(Question)
class QuestionAdmin(admin.ModelAdmin):
    ...
    def get_search_results(self, request, queryset, search_term):
        if search_term:
            queryset = self.model.objects.filter(question_text__icontains=search_term)
        return queryset, True
    ...
```

重启网站，在调查问卷列表页输入"this is"，效果如下图所示。

6.2.7　save_related

方法签名：ModelAdmin.save_related(request, form, formsets, change)

用于保存相关外键对象。4 个参数分别是 HttpRequest、父级 ModelForm 实例、一组 inline formset 对象、用于标记父级模型是添加还是变更的布尔值。

6.2.8 get_autocomplete_fields

方法签名：ModelAdmin.get_autocomplete_fields(request)
返回一组自动完成字段。作用与 autocomplete_fields 一样。

6.2.9 get_readonly_fields

方法签名：ModelAdmin.get_readonly_fields(request, obj=None)
返回一组只读字段。参数 obj 是被编辑对象，当 obj 为 None 时表示新增元素。

6.2.10 get_prepopulated_fields

方法签名：ModelAdmin.get_prepopulated_fields(request, obj=None)
返回一组自动填充字段。作用与 prepopulated_fields 一样。参数 obj 是被编辑对象，当 obj 为 None 时表示新增元素。

6.2.11 get_list_display

方法签名：ModelAdmin.get_list_display(request)
返回一组列表可显示字段。作用与 list_display 一样。

6.2.12 get_list_display_links

方法签名：ModelAdmin.get_list_display_links(request, list_display)
返回一组以超链接形式显示的字段。作用与 list_display_links 一样。

6.2.13 get_exclude

方法签名：ModelAdmin.get_exclude(request, obj=None)
返回一组字段，该组字段与 exclude 属性作用一样。参数 obj 是被编辑对象，当 obj 为 None 时表示新增元素。

6.2.14 get_fields

方法签名：ModelAdmin.get_fields(request, obj=None)
返回一组字段，该组字段与 fields 属性作用一样。参数 obj 是被编辑对象，当 obj 为 None 时表示新增元素。

6.2.15 get_fieldsets

方法签名：ModelAdmin.get_fieldsets(request, obj=None)
返回一组二元元组，与 fieldsets 一样。参数 obj 是被编辑对象，当 obj 为 None 时表示新增元素。

6.2.16 get_list_filter

方法签名：ModelAdmin.get_list_filter(request)
返回一组类似 list_filter 的属性。

6.2.17 get_list_select_related

方法签名:ModelAdmin.get_list_select_related(request)
返回一个布尔值或者一个类似 ModelAdmin.list_select_related 的列表。

6.2.18 get_search_fields

方法签名:ModelAdmin.get_search_fields(request)
返回一组类似 search_fields 的属性。

6.2.19 get_sortable_by

方法签名:ModelAdmin.get_sortable_by(request)
返回一组应用于列表页的可排序字段。
默认情况下,如果设置了 sortable_by 属性,返回 sortable_by,否则返回 get_list_display()。
以下代码将阻止字段 rank 作为可排序字段:

```python
class PersonAdmin(admin.ModelAdmin):

    def get_sortable_by(self, request):
        return {*self.get_list_display(request)} - {'rank'}
```

6.2.20 get_inline_instances

方法签名:ModelAdmin.get_inline_instances(request, obj=None)
返回一组 InlineModelAdmin 对象。
需要注意的是,返回的对象必须是 inlines 实例,否则抛出"Bad Request"异常。

6.2.21 get_inlines

方法签名:ModelAdmin.get_inlines(request, obj)
返回一组可遍历的 inlines 对象。参数 obj 是被编辑对象,当 obj 为 None 时表示新增元素。
通过重写 get_inlines 方法可以为 request 对象或者模型实例动态添加 inlines。

6.2.22 get_urls

方法签名:ModelAdmin.get_urls()
get_urls() 可以用于为 ModelAdmin 设置 URL。
示例代码:

```python
from django.contrib import admin
from django.template.response import TemplateResponse
from django.urls import path

class MyModelAdmin(admin.ModelAdmin):
    def get_urls(self):
        urls = super().get_urls()
        my_urls = [
            path('my_view/', self.my_view),
```

```
        ]
        return my_urls + urls

    def my_view(self, request):
        # ...
        context = dict(
            # Include common variables for rendering the admin template.
            self.admin_site.each_context(request),
            # Anything else you want in the context...
            Key=value,
        )
        return TemplateResponse(request, "sometemplate.html", context)
```

> **注意**
>
> 由于当 Django 路由系统找到第一个匹配的 URL 后就会停止匹配其他 URL，因此最好将自定义 URL 放在默认 URL 之前。

另外需要注意的是，代码中 my_view 没有实现身份认证，同时也没有接收任何 HTTP 请求头信息，无法处理缓存。为了解决这些问题可以使用 AdminSite.admin_view() 对自定义视图进行封装：

```
class MyModelAdmin(admin.ModelAdmin):
    def get_urls(self):
        urls = super().get_urls()
        my_urls = [
            path('my_view/', self.admin_site.admin_view(self.my_view))
        ]
        return my_urls + urls
```

admin_view 方法可以防止未授权的访问，同时为视图添加 django.views.decorators.cache.never_cache() 装饰器，此时启用缓存中间件将不会缓存 my_view。

如果需要使用缓存，可以为 admin_view 方法增加 cacheable 参数：

```
path('my_view/', self.admin_site.admin_view(self.my_view, cacheable=True))
```

6.2.23　get_form

方法签名：ModelAdmin.get_form(request, obj=None, **kwargs)

为新增和编辑页返回 ModelForm。

示例代码：

```
class MyModelAdmin(admin.ModelAdmin):
    def get_form(self, request, obj=None, **kwargs):
        if request.user.is_superuser:
            kwargs['form'] = MySuperuserForm
        return super().get_form(request, obj, **kwargs)
```

6.2.24　get_formsets_with_inlines

方法签名：ModelAdmin.get_formsets_with_inlines(request, obj=None)

用生成器（yield）在管理后台的添加和编辑页面返回一组 (FormSet, InlineModelAdmin)。

例如，以下代码可以为编辑页面返回部分特定的 inline 对象：

```
class MyModelAdmin(admin.ModelAdmin):
    inlines = [MyInline, SomeOtherInline]

    def get_formsets_with_inlines(self, request, obj=None):
        for inline in self.get_inline_instances(request, obj):
            # hide MyInline in the add view
            if not isinstance(inline, MyInline) or obj is not None:
                yield inline.get_formset(request, obj), inline
```

6.2.25　formfield_for_foreignKey

方法签名：ModelAdmin.formfield_for_foreignKey(db_field, request, **kwargs)

formfield_for_foreignKey() 方法允许开发人员重写外键对应的 formfield。默认情况下参数 db_field 是外键，formfield_for_foreignKey() 将会取得外键对应的表单字段。

例如，基于当前用户返回对象的子集：

```
class MyModelAdmin(admin.ModelAdmin):
    def formfield_for_foreignKey(self, db_field, request, **kwargs):
        if db_field.name == "car":
            kwargs["queryset"] = Car.objects.filter(owner=request.user)
        return super().formfield_for_foreignKey(db_field, request, **kwargs)
```

6.2.26　formfield_for_manytomany

方法签名：ModelAdmin.formfield_for_manytomany(db_field, request, **kwargs)

与 formfield_for_foreignKey() 方法相似，formfield_for_manytomany() 可以用于重写多对多字段对应的表单字段。

当一辆车可以属于多个人且一个人可以拥有多辆车时，可以按如下方式修改代码：

```
class MyModelAdmin(admin.ModelAdmin):
    def formfield_for_manytomany(self, db_field, request, **kwargs):
        if db_field.name == "cars":
            kwargs["queryset"] = Car.objects.filter(owner=request.user)
        return super().formfield_for_manytomany(db_field, request, **kwargs)
```

6.2.27　formfield_for_choice_field

方法签名：ModelAdmin.formfield_for_choice_field(db_field, request, **kwargs)

与 formfield_for_foreignKey() 方法、formfield_for_manytomany() 方法相似，formfield_for_choice_field() 可以用于重写 choices 字段对应的表单字段。

例如，当超级用户与普通用户拥有不同的 choices 字段时，可以按照如下方式编写代码：

```
class MyModelAdmin(admin.ModelAdmin):
    def formfield_for_choice_field(self, db_field, request, **kwargs):
        if db_field.name == "status":
            kwargs['choices'] = (
                ('accepted', 'Accepted'),
                ('denied', 'Denied'),
```

```
            )
            if request.user.is_superuser:
                kwargs['choices'] += (('ready', 'Ready for deployment'),)
        return super().formfield_for_choice_field(db_field, request, **kwargs)
```

6.2.28　get_changelist

方法签名：ModelAdmin.get_changelist(request, **kwargs)

为列表显示返回 ChangeList 类。默认情况下返回 django.contrib.admin.views.main.ChangeList。

6.2.29　get_changelist_form

方法签名：ModelAdmin.get_changelist_form(request, **kwargs)

返回列表页对应的 ModelForm 类。

示例代码：

```
from django import forms

class MyForm(forms.ModelForm):
    pass

class MyModelAdmin(admin.ModelAdmin):
    def get_changelist_form(self, request, **kwargs):
        return MyForm
```

6.2.30　get_changelist_formset

方法签名：ModelAdmin.get_changelist_formset(request, **kwargs)

当启用 list_editable 时，返回列表页对应的 ModelFormSet 类。

示例代码：

```
from django.forms import BaseModelFormSet

class MyAdminFormSet(BaseModelFormSet):
    pass

class MyModelAdmin(admin.ModelAdmin):
    def get_changelist_formset(self, request, **kwargs):
        kwargs['formset'] = MyAdminFormSet
        return super().get_changelist_formset(request, **kwargs)
```

6.2.31　lookup_allowed

方法签名：ModelAdmin.lookup_allowed(lookup, value)

通过用户请求的 URL 中的查询字符串可以得到 lookup。lookup 可用于过滤列表页的显示内容，这就是 list_filter 的工作原理。Django 曾经存在一个安全缺陷，当用户在 URL 中输入任意内容后，Django 没有安全检查就直接解析用户请求内容，这导致用户可以轻松地跳过安全检查而访问敏感数据，为此 Django 在 ModelAdmin 中增加了 lookup_allowed 方法。

lookup_allowed 方法从 URL 中取得查询字符串以及对应的值，如果 URL 中的查询字符串是

user__email=user@example.com，那么查询字符串是 user__email，对应的值是 user@example.com，最后 lookup_allowed 方法返回一个布尔值用于标志是否有权限使用当前查询字符串过滤 QuerySet；如果返回 False，那么意味着用户非法访问资源，这将导致 Django 抛出 DisallowedModelAdminLookup 异常（SuspiciousOperation 的子类）。

默认情况下，lookup_allowed 方法允许访问模型的本地字段，查询路径来自于 list_filter（注意不是 get_list_filter()）。另外，lookup 要能正确匹配 ForeignKey.limit_choices_to。

6.2.32 has_view_permission

方法签名：ModelAdmin.has_view_permission(request, obj=None)

标志是否允许用户访问对象（obj），允许访问时返回 True，否则返回 False。如果对象是 None，一般也需要返回 True 或 False 来表示用户是否有权限访问该类型的对象（False 表示不允许当前用户访问任何该类型的对象）。

默认情况下，如果用户拥有 change 和 view 的权限，has_view_permission() 方法返回 True。

6.2.33 has_add_permission

方法签名：ModelAdmin.has_add_permission(request)

如果用户拥有 add 权限则返回 True，否则返回 False。

6.2.34 has_change_permission

方法签名：ModelAdmin.has_change_permission(request, obj=None)

如果用户拥有 change 权限则返回 True，否则返回 False。如果对象是 None，一般也需要返回 True 或 False 来表示用户是否有权限编辑该类型的对象（False 表示不允许当前用户编辑任何该类型的对象）。

6.2.35 has_delete_permission

方法签名：ModelAdmin.has_delete_permission(request, obj=None)

如果用户拥有 delete 权限则返回 True，否则返回 False。如果对象是 None，一般也需要返回 True 或 False 来表示用户是否有权限删除该类型的对象（False 表示不允许当前用户删除任何该类型的对象）。

6.2.36 has_module_permission

方法签名：ModelAdmin.has_module_permission(request)

如果在后台管理首页显示 module 并且允许用户访问 module 首页则返回 True，否则返回 False。重写该方法不能改变对 view、add、change、delete 视图的访问权限，如果需要改变对以上视图的访问权限，请重写 has_view_permission()、has_add_permission()、has_change_permission() 和 has_delete_permission() 方法。

6.2.37 get_queryset

方法签名：ModelAdmin.get_queryset(request)

get_queryset() 方法返回一个 QuerySet 对象，QuerySet 所包含的全部模型实例允许在管理后台编辑。例如，下面代码能够根据当前登录用户返回模型实例：

```python
class MyModelAdmin(admin.ModelAdmin):
    def get_queryset(self, request):
        qs = super().get_queryset(request)
        if request.user.is_superuser:
            return qs
        return qs.filter(author=request.user)
```

6.2.38　message_user

方法签名：ModelAdmin.message_user(request, message, level=messages.INFO, extra_tags='', fail_silently=False)

使用 django.contrib.messages 引擎向用户发送消息。

方法参数允许变更邮件级别，添加额外的 CSS 标签。如果 fail_silently=True，那么当没有安装 contrib.messages 时，只会安静失败。

6.2.39　get_paginator

方法签名：ModelAdmin.get_paginator(request, queryset, per_page, orphans=0, allow_empty_first_page=True)

为当前视图返回一个分页器实例。

6.2.40　response_add

方法签名：ModelAdmin.response_add(request, obj, post_url_continue=None)

决定了 add_view() 的 HttpResponse 对象。

当后台管理表单提交之后和全部相关对象创建并保存成功时，Django 调用 response_add() 方法。重写该方法可以改变对象保存成功后的行为。

6.2.41　response_change

方法签名：ModelAdmin.response_change(request, obj)

决定了 change_view () 的 HttpResponse 对象。

当后台管理表单提交之后并且全部相关对象保存成功时，Django 调用 response_change() 方法。重写该方法可以改变对象保存成功后的行为。

6.2.42　response_delete

方法签名：ModelAdmin.response_delete(request, obj_display, obj_id)

决定了 delete_view () 的 HttpResponse 对象。

当对象删除成功后，调用 response_delete() 方法。重写该方法可以改变对象删除成功后的行为。

obj_display 是被删除对象的名字字符串。

obj_id 是被删除对象的序列化标志符。

6.2.43 get_changeform_initial_data

方法签名：ModelAdmin.get_changeform_initial_data(request)

用于初始化新增页面（注意实际使用中是新增页面）数据的钩子。默认情况下，表单字段值取自 GET 参数。例如，当 URL 包含 ?name=initial_value 时，get_changeform_initial_data() 方法将会给表单参数 name 赋值为 initial_value。使用该方法可以返回一个字典作为初始化编辑页面的数据源，例如，以下方法将会使问卷选项编辑页面的 choice_text 字段默认填充文字"测试数据"：

```
@admin.register(Choice)
class ChoiceAdmin(admin.ModelAdmin):
    ...
    def get_changeform_initial_data(self, request):
        data = super().get_changeform_initial_data(request)
        data["choice_text"] = data.get("choice_text","请输入问卷选项 ")
        return data
    ...
```

此时添加问卷选项的页面默认显示效果如下图所示。

修改 URL 如下：

http://127.0.0.1:8000/admin/polls/choice/add/?choice_text= 你好，Django!

刷新页面，效果如下图所示。

6.2.44 get_deleted_objects

方法签名：ModelAdmin.get_deleted_objects(objs, request)

用于自定义删除对象的过程以及选择被删除对象的行为。

参数 objs 是被删除对象，可以是一个可遍历的 QuerySet，也可以是一个模型实例的列表。参数 request 是 HttpRequest 对象。

该方法必须返回一个包含 4 个元素的元组，4 个元素分别如下。

deleted_objects：是一组表示被删除对象名字的字符串。如果在删除对象的同时有其他关联对象被删除，那么 deleted_objects 应该是一个嵌套的数组，该嵌套数组与 HTML 中的无序列表非常相似，唯一的区别是不包含 <ul> 标签，例如：

```
<li>States
<ul>
        <li>Kansas
        <ul>
                <li>Lawrence</li>
                <li>Topeka</li>
        </ul>
```

```
            </li>
            <li>Illinois</li>
    </ul>
</li>
```

model_count：字典类型，字典的键是模型的 verbose_name_plural，对应的值是被删除的对象数量。

perms_needed：一组用户无权删除的模型的 verbose_name。

protected：一组字符串表示不能被删除的相关联的对象，这个列表将会显示在模板中。

6.2.45 add_view

方法签名：ModelAdmin.add_view(request, form_url='', extra_context=None)

添加模型实例的视图。

6.2.46 change_view

方法签名：ModelAdmin.change_view(request, object_id, form_url='', extra_context=None)

编辑模型实例的视图。

6.2.47 changelist_view

方法签名：ModelAdmin.changelist_view(request, extra_context=None)

模型列表视图。

6.2.48 delete_view

方法签名：ModelAdmin.delete_view(request, object_id, extra_context=None)

确认删除模型实例的视图。

6.2.49 history_view

方法签名：ModelAdmin.history_view(request, object_id, extra_context=None)

特定模型修改历史的视图。

6.3 ModelAdmin 资源

Django 允许用户在 ModelAdmin 类中通过增加一个内部 Media 类的方式为 add 和 change 视图添加额外的 CSS 或者 JavaScript 的能力，例如：

```
class ArticleAdmin(admin.ModelAdmin):
    class Media:
        css = {
            "all": ("my_styles.css",)
        }
        js = ("my_code.js",)
```

这里的静态文件基于 STATIC_URL（如果 STATIC_URL 是 None，则基于 MEDIA_URL）。

6.3.1 jQuery

Django 管理后台为用户提供对 jQuery 的支持。

为了避免和用户提供的 jQuery 冲突，Django 所提供的 jQuery 的命名空间是 django.jQuery，Django3.0 提供的 jQuery 版本是 3.4.1。

Django 同时提供了 jQuery 的开发版和压缩版。当 DEBUG=True 时，将会使用开发版（jquery.js），否则使用压缩版（jquery.min.js）。

6.4 定制验证功能

在 Django 管理后台添加自定义数据验证功能非常容易，只要重写 ModelAdmin 的 form 属性即可，例如：

```
class ArticleAdmin(admin.ModelAdmin):
    form = MyArticleAdminForm

class MyArticleAdminForm(forms.ModelForm):
    def clean_name(self):
        # do something that validates your data
        return self.cleaned_data["name"]
```

6.5 InlineModelAdmin

使用 Django 管理后台，可以在编辑上级模型对象的同时编辑关联对象，这些关联对象就叫作 inline。例如，前面添加调查问卷时可以同时添加问卷选项，问卷选项就是问卷的 inline。

InlineModelAdmin 有两个子类：TabularInline 和 StackedInline。

这两个子类的区别仅仅是显示数据时所使用的模板不同。

InlineModelAdmin 提供了很多 ModelAdmin 的功能，如：

- form
- fieldsets
- fields
- formfield_overrides
- exclude
- filter_horizontal
- filter_vertical
- ordering
- prepopulated_fields
- get_fieldsets()
- get_queryset()
- radio_fields
- readonly_fields
- raw_id_fields
- formfield_for_choice_field()

- formfield_for_foreignKey()
- formfield_for_manytomany()
- has_module_permission()

除上所述这些，InlineModelAdmin 额外提供或者自定义了其他属性和方法，接下来一一介绍。

6.5.1　InlineModelAdmin.model

必选属性，inline 所使用的模型。

6.5.2　InlineModelAdmin.fk_name

模型的外键。大多数情况下外键都是自动处理的，但是，如果模型存在多个外键时，则必须显式地给出 fk_name。

6.5.3　InlineModelAdmin.formset

默认值是 BaseInlineFormSet。使用定制化的 formset 可以使得程序有更多的可能性。

6.5.4　InlineModelAdmin.form

默认值是 ModelForm。为 inline 创建 formset 时传递给 inlineformset_factory() 的 form 对象。

6.5.5　InlineModelAdmin.classes

一组 CSS 类，这些类在渲染 inline 对象时将被应用于 fieldset。默认值为 None。下面以 collapse 类为例演示该属性的显示效果。

修改 ChoiceInline 类：

```
class ChoiceInline(admin.StackedInline):
    model = Choice
    classes = ('collapse', )
```

重启 Django 应用，在管理后台增加一个问卷，此时页面显示效果如下图所示。

此时 CHOICES 默认为折叠状态，单击"显示"按钮可以展开 CHOICES。

6.5.6　InlineModelAdmin.extra

该属性用于控制初始化 inline 对象时默认显示几个表单。例如，添加问卷时默认可以添加 3 个问卷选项，如果将 extra 属性设置为 0，则默认不显示任何选项表单，如下图所示。

如果将 extra 属性设置为 1，则显示一个选项表单，如下图所示。

如果需要添加更多问卷选项，可以单击链接"添加另一个 Choice"。当页面中显示的问卷选项表单数大于 max_num 时或者禁用 JavaScript 时，不显示链接"添加另一个 Choice"。

开发人员可以重写方法 InlineModelAdmin.get_extra() 来自定义 max_num 的值。

6.5.7 InlineModelAdmin.max_num

用于控制 inline 对象可以显示的最大表单数。
InlineModelAdmin.get_max_num() 同样可以改变最大表单数。

6.5.8 InlineModelAdmin.min_num

用于控制 inline 对象可以显示的最小表单数。
InlineModelAdmin.get_min_num() 同样可以改变最小表单数。

6.5.9　InlineModelAdmin.raw_id_fields

默认情况下，Django 管理后台使用下拉列表的形式处理外键，但是，当用户充分了解外键值的情况下，只须设置对应的外键值即可，而不用再在一长串下拉列表中查找外键了。

使用方式如下：

```
raw_id_fields = ('question',)
```

当对 ManyToManyField 属性使用 raw_id_fields 时可能会出现以下错误：

```
ERRORS:
<class 'polls.admin.QuestionInline'>: (admin.E002) The value of 'raw_id_fields[0]'
refers to 'question', which is not an attribute of 'polls.Question'.
<class 'polls.admin.QuestionInline'>: (admin.E202) 'polls.Question' has no
ForeignKey to 'polls.Category'.
```

此时可以使用 through 属性解决异常，through 是由两个外键组成的新模型，以 Category 和 Question 类为例修改代码：

```
class QuestionInline(admin.StackedInline):
    model = Category.question_name.through
    raw_id_fields = ('question',)

@admin.register(Category)
class CategoryAdmin(admin.ModelAdmin):
    inlines = (QuestionInline,)
```

重启 Django 应用，添加一个新的问卷分类，此时可以通过输入问卷 ID 的方式来完成问卷分类的增加，效果如下图所示。

观察以上网页截图可以发现，页面上同时有两处地方可以添加问卷：Question name 和 CATEGORY-QUESTION 关系。为了避免冲突，还需要在 CategoryAdmin 中排除 question：

```
@admin.register(Category)
class CategoryAdmin(admin.ModelAdmin):
    filter_horizontal = ('question_name',)
    inlines = (QuestionInline,)
    exclude = ('question_name',)
    pass
```

6.5.10　InlineModelAdmin.template

用于渲染 inline 对象的模板。

6.5.11　InlineModelAdmin.verbose_name

用于重写模型内部 Meta 类的 verbose_name。

6.5.12　InlineModelAdmin.verbose_name_plural

用于重写模型内部 Meta 类的 verbose_name_plural。

6.5.13　InlineModelAdmin.can_delete

配置 inline 对象是否可以被删除。默认值为 True。

当 can_delete = False 时，编辑调查问卷时将不能删除问卷选项，如下图所示。

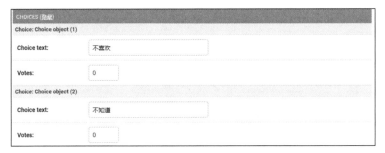

而默认时，选择后包含删除复选框，如下图所示。

6.5.14 InlineModelAdmin.show_change_link

在编辑模型页面的 inline 对象旁边配置一个链接，当单击这个链接时可以进入编辑 inline 对象的页面。

例如，修改 ChoiceInline 类，允许用户在编辑调查问卷时编辑 inline 对象：

```
class ChoiceInline(admin.StackedInline):
    ...
    show_change_link = True
```

重启 Django 项目，进入调查问卷编辑页面，如下图所示。

6.5.15 InlineModelAdmin.get_formset(request, obj=None, **kwargs)

该方法返回一个 BaseInlineFormSet 类，在 Django 管理后台添加或者编辑模型时使用该方法的返回值。编辑模型时，参数 obj 是被编辑 inline 的父对象，如编辑调查问卷"你喜欢 Django 吗？"时，obj 内容如下图所示。

当添加 inline 的父对象时，参数 obj 是 None。

6.5.16 InlineModelAdmin.get_extra(request, obj=None, **kwargs)

该方法返回额外使用的 inline 表单数量。默认返回 InlineModelAdmin.extra 属性。

重写该方法可以以编程的方式改变 extra 值，例如：

```
class BinaryTreeAdmin(admin.TabularInline):
    model = BinaryTree

    def get_extra(self, request, obj=None, **kwargs):
        extra = 2
        if obj:
            return extra - obj.binarytree_set.count()
        return extra
```

6.5.17 InlineModelAdmin.get_max_num(request, obj=None, **kwargs)

该方法返回允许使用的最大 inline 表单数。默认返回 InlineModelAdmin.max_num 属性。

重写该方法可以以编程的方式改变 max_num 值,例如:

```
class BinaryTreeAdmin(admin.TabularInline):
    model = BinaryTree

    def get_max_num(self, request, obj=None, **kwargs):
        max_num = 10
        if obj and obj.parent:
            return max_num - 5
        return max_num
```

6.5.18　InlineModelAdmin.get_min_num(request, obj=None, **kwargs)

该方法返回允许使用的最小 inline 表单数。默认返回 InlineModelAdmin.min_num 属性。

6.5.19　InlineModelAdmin.has_add_permission(request, obj)

如果允许用户添加 inline 对象则返回 True,否则返回 False。

6.5.20　InlineModelAdmin.has_change_permission(request, obj=None)

如果允许用户编辑 inline 对象则返回 True,否则返回 False。

6.5.21　InlineModelAdmin.has_delete_permission(request, obj=None)

如果允许用户删除 inline 对象则返回 True,否则返回 False。

6.5.22　使用中间模型处理 ManyToMany 关系

在 6.5.9 节介绍 raw_id_fields 时,使用 through 属性来处理 ManyToManyField,through 属性会带入一个中间模型,但是 Django 管理后台并不会显示它。

如果想要编辑中间模型,那么就需要在管理后台显示它,此时可以通过增加一个中间模型的方式来达到相应目的。这里,仍然以调查问卷和问卷分类为例,首先增加一个中间模型:

```
class Membership(models.Model):
    question = models.ForeignKey(Question, on_delete=models.CASCADE)
    category = models.ForeignKey(Category, on_delete=models.CASCADE)
```

接下来编写 MembershipInline:

```
from .models import Membership
class MembershipInline(admin.TabularInline):
    model = Membership
    extra = 1
```

最后修改 QuestionAdmin 和 CategoryAdmin:

```
@admin.register(Question)
class QuestionAdmin(admin.ModelAdmin):
    ...
    inlines = [ChoiceInline, MembershipInline]

@admin.register(Category)
```

```
class CategoryAdmin(admin.ModelAdmin):
    inlines = (MembershipInline,)
    exclude = ('question',)
```

因为增加了新的模型，所以在重启 Django 应用程序前，首先执行数据库迁移命令使模型生效。输入命令，如下图所示。

最后刷新管理后台，添加一个新的问卷分类，效果如下图所示。

6.6 重写管理后台模板

开发人员可以很容易地重写 Django 管理后台页面所使用的模板。可以针对具体应用程序重写模板，也可以为指定模型重写模板，本节介绍如何重写管理后台模板。

6.6.1 新建管理后台模板

Django 管理后台的模板文件默认保存在 Python 安装目录下的 contrib/admin/templates/admin 文件夹中。为了重写模板文件，需要在工程的 templates 文件夹下新建一个 admin 文件夹。如果在配置文件中修改了 TEMPLATES 的 DIRS，那么 admin 文件夹可以是 DIRS 中的任意子文件夹。

为了区分新模板属于哪个应用程序，在 admin 文件夹下以应用程序名新建子文件夹，如 polls。

注意

由于 Django 在查找文件路径时会将模型名字转换为小写字母，因此，新建任何文件夹时都使用小写字母。

如果只是希望重写现有模板,那么最简单的方式是将对应的原始模板文件复制过来,直接编辑。

以列表页为例,如果希望对列表的展示样式做一些美化,可以将 change_list.html 复制到新路径,复制后的主要目录结构如下:

```
polls/
    __init__.py
    admin.py
apps.py
    migrations/
    statics/
    templates/
        admin/
            polls/
                change_list.html
        polls/
    models.py
    tests.py
    views.py
```

如果只希望修改模型 Category 的列表页样式,可以在 polls 下新建一个文件夹 category,将 change_list.html 复制到 category,新的文件结构如下:

```
polls/
    __init__.py
    admin.py
    apps.py
    migrations/
    statics/
    templates/
        admin/
            polls/
                category/
                    change_list.html
                change_list.html
            polls/
    models.py
    tests.py
    views.py
```

修改 category/change_list.html 数据展示代码,使得在没有任何问卷分类信息的时候能够展示一段提示性文字。

在代码底部找到 block result_list,替换为以下代码:

```
{% block result_list %}
  {% if cl.result_list or cl.formset %}
    {% result_list cl %}
  {% else %}
      <p>暂时还没有任何问卷分类</p>
  {% endif %}
{% endblock %}
```

代码分析:

变量 cl 是 ChangeList 类的实例,用于保存完整的页面状态。

result_list 是 contrib/admin/templatetags/admin_list.py 中定义的一个模板标签，用于显示表头和列表内容：

```python
def result_list(cl):
    """
    Display the headers and data list together.
    """
    headers = list(result_headers(cl))
    num_sorted_fields = 0
    for h in headers:
        if h['sortable'] and h['sorted']:
            num_sorted_fields += 1
    return {
        'cl': cl,
        'result_hidden_fields': list(result_hidden_fields(cl)),
        'result_headers': headers,
        'num_sorted_fields': num_sorted_fields,
        'results': list(results(cl)),
    }
```

当问卷分类页没有数据时，页面显示效果如下图所示。

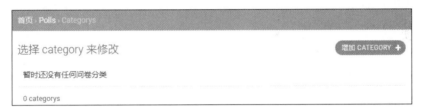

增加一条分类信息后刷新页面，此时与 Django 的默认列表页显示一样，如下图所示。

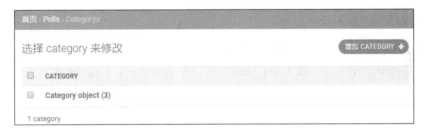

6.6.2 重写与替换

在实际工作中，到底是应该选择重写模板文件还是替换模板文件，需要根据实际情况来决定。由于 Django 是模块化开发的，因此，通常不建议开发人员像替换 change_list.html 文件一样完整地替换模板。往往在实际工作中只需要通过重写模板的部分 block 即可，例如：

```
{% extends "admin/polls/change_list.html" %}
{% load i18n admin_urls static admin_list %}

{% block result_list %}
  {% if cl.result_list or cl.formset %}
    {% result_list cl %}
```

```
    {% else %}
      <p>暂时还没有任何问卷分类</p>
    {% endif %}
{% endblock %}
```

6.6.3 可重写模板

需要注意的是,只有以下 Django 管理后台模板能被重写:
- actions.html
- app_index.html
- change_form.html
- change_form_object_tools.html
- change_list.html
- change_list_object_tools.html
- change_list_results.html
- date_hierarchy.html
- delete_confirmation.html
- object_history.html
- pagination.html
- popup_response.html
- prepopulated_fields_js.html
- search_form.html
- submit_line.html

对于不能被重写的模板,仍然可以通过替换的方式实现模板更新,尤其是 400 和 500 错误页。

6.6.4 根模板和登录模板

如果希望改变 Django 应用程序的 index、login、logout 模板,最好使用自己开发的 AdminSite 实例,在 AdminSite 实例中修改 AdminSite.index_template、AdminSite.login_template 或者 AdminSite.logout_template 属性。

6.7 AdminSite

6.7.1 重写 AdminSite

Django 自带的管理后台是 django.contrib.admin.sites.AdminSite 的一个实例,对高级开发人员来说,可能无法满足自身需求,此时可以通过重写 AdminSite 的方式,对默认管理后台进行扩展。

首先重写 polls\admin.py 文件,编写新的 MyAdminSite 类,用它替代默认的 AdminSite:

```
from django.contrib.admin import AdminSite
from .models import Question, Choice, Category, Membership

import os
```

```python
class MyAdminSite(AdminSite):
    site_header = 'My New Administration Page'

admin_site = MyAdminSite(name='myadmin')

admin_site.register(Question)
admin_site.register(Choice)
admin_site.register(Category)
admin_site.register(Membership)
```

其次重写 polls\apps.py,使 PollsConfig 指向前面编写的 MyAdminSite 类:

```python
from django.contrib.admin.apps import AdminConfig

class PollsConfig(AdminConfig):
    name = 'polls'
    default_site = 'polls.admin.MyAdminSite'

    def ready(self):
        from django.contrib import admin
        from django.contrib.admin import sites

        class MyAdminSite(admin.AdminSite):
            pass

        mysite = MyAdminSite()
        admin.site = mysite
        sites.site = mysite
```

再次修改 mysite\urls.py,使新的 URL 指向 MyAdminSite 的实例:

```python
from polls.admin import admin_site
from django.urls import include, path

urlpatterns = [
    path('polls/', include('polls.urls')),
    path('myadmin/', admin_site.urls),
]
```

最后在 settings.py 中注册新的 AdminSite:

```python
INSTALLED_APPS = [
    'django.contrib.admin.apps.SimpleAdminConfig',
    'django.contrib.auth',
    'django.contrib.contenttypes',
    'django.contrib.sessions',
    'django.contrib.messages',
    'django.contrib.staticfiles',
    'polls.apps.PollsConfig',
    'django.contrib.humanize',
]
```

因为默认的 admin 类会调用 autodiscovery 方法查找并注册默认的管理后台模块,所以如果项目不需要使用这些模块,可以使用 'django.contrib.admin.apps.SimpleAdminConfig' 替代默认的 'django.contrib.admin'。

新的 Django 管理后台显示效果如下图所示。

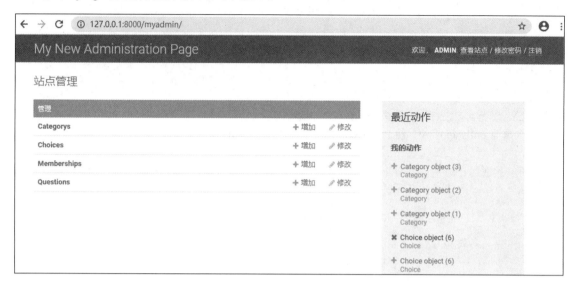

6.7.2 多管理后台的实现

前面实现了一个自定义管理后台，在此基础上可以无限增加新的管理后台，开发人员只要继续编写 AdminSite 实例并将新的实例引用到 URL 即可。下面是一个包含 'basic-admin/' 和 'advanced-admin/' 两个管理后台的 URLs：

```
# urls.py
from django.urls import path
from myproject.admin import advanced_site, basic_site

urlpatterns = [
    path('basic-admin/', basic_site.urls),
    path('advanced-admin/', advanced_site.urls),
]
```

注意

为了讲解其他内容，在完成本章后请恢复代码，删除自定义管理后台相关内容。

第 7 章 路由系统

简洁优雅的 URL 是高质量 Web 应用程序的象征。Django 允许开发人员设计任何形式的 URL，这在早期的网站中是不可想象的。早期的网站通常会使用很长的一串 URL，而且会包括一些无用信息，如 .aspx、.php 等。

为了给应用程序设计 URL，开发人员需要开发一个 Python 模块，这个模块就是 URL 的配置信息，通常将这个配置模块叫作 URLconf。这个模块是一个纯粹的 Python 脚本，它包含了 URL 表达式与 Python 方法之间的映射，这里的 Python 方法就是 Django 应用程序中的视图方法。第 3 章的示例中的 mysite/urls.py 和 polls/urls.py 就是两个 URLconf 实例。

7.1 Django 处理 HTTP 请求的流程

当用户发起一个 HTTP 请求时，Django 会按照以下逻辑对请求进行处理。

（1）确定 URLconf 的根配置位置，通常 URL 根配置在 ROOT_URLCONF 中设置。如果请求的 HttpRequest 对象包含 urlconf 属性，则按照 urlconf 属性查找 URLconf。

（2）加载配置信息，在配置信息中查找 urlpatterns。

（3）按顺序检索 urlpatterns 中的所有 URL 模式字符串，并定位在第一个与 URL 匹配的 URL 模式字符串。

（4）当检索到匹配的 URL 模式字符串后，调用对应的视图方法，并传递以下参数给视图方法：
- ❑ 一个 HttpRequest 对象实例。
- ❑ 如果匹配的 URL 模式字符串包含未命名的组，那么匹配的信息会作为位置参数传递给视图。
- ❑ 路径表达式中的命名部分组成了视图的关键字参数，以上关键字参数会被 django.urls.path() 或者 django.urls.re_path() 的 kwargs 覆盖。

（5）如果在 URLconf 中没有找到任何匹配的模式字符串，或者出现其他任何错误，Django 将会调用一个用于处理错误信息的视图。

7.2 URLconf 示例

下面是一个简单的 URLconf 代码示例：

```
from django.urls import path

from . import views
```

```
urlpatterns = [
    path('articles/2003/', views.special_case_2003),
    path('articles/<int:year>/', views.year_archive),
    path('articles/<int:year>/<int:month>/', views.month_archive),
    path('articles/<int:year>/<int:month>/<slug:slug>/', views.article_detail),
]
```

代码分析：
- 函数 path 的第一个参数是一个 URL 模式字符串，用于匹配 URL，类似正则表达式。
- 函数 path 的第二个参数是用于处理 URL 请求的视图函数。
- 使用尖括号提取 URL 中的参数，如 <int:year>。
- 使用类型转化器对参数类型进行转换，如 int: 会将从 URL 中捕获的值转换为数值类型，如果没有指定类型转换器，如 <year>，则任何不包含斜杠（/）的字符串都会被提取。
- URL 模式字符串不需要以 / 开头。

应用场景：
- 发送向 /articles/2005/03/ 的请求将会与第三个 URL 模式字符串匹配成功，匹配成功后调用 views.month_archive(request, year=2005, month=3)。
- 发送向 /articles/2003/ 的请求将会与第一个 URL 模式字符串匹配成功而不是第二个，因为 Django 在第一次匹配成功后停止后续 URL 检验，本例匹配成功后调用 views.special_case_2003(request)。
- 发送向 /articles/2003 的请求不会与任何 URL 模式字符串匹配成功，因为每个 URL 模式字符串都要求以 / 结尾。
- 发送向 /articles/2003/03/building-a-django-site/ 的请求将会与最后一个 URL 模式字符串匹配成功，匹配成功后调用 views.article_detail(request, year=2003, month=3, slug= " building-a-django-site ")。

7.3 URL 参数类型转化器

7.2 节提到可以使用 int 对捕捉到的 URL 参数进行类型转换，Django 支持的所有类型转换器如下。
- str：匹配任意非空字符串，但是不能匹配 URL 分隔符"/"。这是默认的 URL 参数转换器。
- int：匹配任意大于等于 0 的整数。
- slug 匹配任意 slug 字符串，slug 字符串可以包含 ASCII 字母、数字、连字符"-"和下画线"_"。
- uuid：匹配 UUID 字符串（字符串中的字母必须为小写），例如 075194d3-6885-417e-a8a8-6c931e272f00。
- path：匹配任意非空字符串，包括 URL 分隔符斜杠（/）。它允许匹配完整的 URL 而不是 URL 的一个片段。

7.4 自定义 URL 参数类型转化器

对于更加复杂的场景，开发人员可以开发自定义参数类型转换器。自定义参数类型转换器包括以下部分：

- 一个 regex 属性，属性值为正则表达式。
- 一个 to_python(self, value) 方法，该方法用于将匹配的 URL 参数转换为指定类型，当类型转换失败后抛出 ValueError 异常。
- 一个 to_url(self, value) 方法，该方法用于将 Python 类型转换为类型转换器字符串。

下面是一个用于捕获日期年的类型转换器：

```python
class FourDigitYearConverter:
    regex = '[0-9]{4}'

    def to_python(self, value):
        return int(value)

    def to_url(self, value):
        return '%04d' % value
```

使用 register_converter() 将以上类型转换器注册到 URLconf：

```python
#!/usr/bin/python
# -*- coding: UTF-8 -*-

from django.urls import register_converter, path

from . import views
from .converters.FourDigitYearConverter import *

register_converter(FourDigitYearConverter, 'yyyy')

app_name = 'polls'
urlpatterns = [
    path('<yyyy:year>/', views.get_year, name='detail'),
]
```

创建测试视图：

```python
def get_year(request, year):
    return HttpResponse(str(year))
```

此时目录结构如下：

```
polls/
    __init__.py
    admin.py
    apps.py
    converters/
        __init__.py
        FourDigitYearConverter.py
    migrations/
    statics/
    templates/
    models.py
    tests.py
    views.py
```

启动 Web 服务，访问 URL，效果如下图所示。

7.5 使用正则表达式

与早期的 Django 一样，Django 3.0 仍然可以使用正则表达式匹配 URL，此时需要使用 re_path() 方法而不是 path()。

Python 的正则表达式支持对分组进行命名，语法格式为 (?P<name>pattern)，其中，name 为分组名，pattern 为匹配的正则表达式。

使用正则表达式对前面的 URLconf 进行重写：

```
from django.urls import path, re_path

from . import views

urlpatterns = [
    path('articles/2003/', views.special_case_2003),
    re_path(r'^articles/(?P<year>[0-9]{4})/$', views.year_archive),
    re_path(r'^articles/(?P<year>[0-9]{4})/(?P<month>[0-9]{2})/$', views.month_archive),
    re_path(r'^articles/(?P<year>[0-9]{4})/(?P<month>[0-9]{2})/(?P<slug>[\w-]+)/$',
views.article_detail),
]
```

虽然可以使用未命名的正则表达式，例如使用 ([0-9]{4}) 替代 (?P<year>[0-9]{4})，但是为了防止出现意外错误，推荐对分组进行命名。

需要注意的是，不要将命名正则表达式与未命名正则表达式混合使用，这样会造成未命名正则表达式丢失。

正则表达式可以嵌套使用，如：

```
re_path(r'comments/(?:page-(?P<page_number>\d+)/)?$', comments)
```

7.6 导入其他 URLconf

对现代 Web 应用程序来说，一个工程通常会包含多个应用程序，每个应用程序包含很多 URL，如果将这些 URL 都写在 URLconf 根模块中，那么 URLconf 将会变得非常臃肿，不利于维护。对于这种情况，常用的解决办法就是为每一个应用程序写一套独立的 URLconf，而 URLconf 根模块通过使用 include() 方法将其他 URLconf 引用进来。

下面是 Polls 网站的 mysite/urls.py：

```
from django.contrib import admin
from django.urls import include, path

urlpatterns = [
```

```
    path('polls/', include('polls.urls')),
    path('admin/', admin.site.urls),
]
```

当 Django 遇到 include() 方法时，匹配工作跳转到被引用的 URLconf 进行验证。

使用 include() 方法还可以引用其他 URL 模式列表，例如：

```
from django.urls import include, path

from apps.main import views as main_views
from credit import views as credit_views

extra_patterns = [
    path('reports/', credit_views.report),
    path('reports/<int:id>/', credit_views.report),
    path('charge/', credit_views.charge),
]

urlpatterns = [
    path('', main_views.homepage),
    path('help/', include('apps.help.urls')),
    path('credit/', include(extra_patterns)),
]
```

此时访问 /credit/reports/ 将会调用 credit_views.report() 视图方法。这样做的好处是，当一个应用程序中多条 URL 的前缀相同时，如在本例中 extra_patterns 中的 URL 都是以 credit 开头，可以简化 URL 模式字符串。

最后，当多个 url 模式字符串拥有相同的前缀时还可以对 url 前缀进行提取，例如：

```
urlpatterns = [
    path('<page_slug>-<page_id>/history/', views.history),
    path('<page_slug>-<page_id>/edit/', views.edit),
    path('<page_slug>-<page_id>/discuss/', views.discuss),
    path('<page_slug>-<page_id>/permissions/', views.permissions),
]
```

前面这些 url 都包含 '<page_slug>-<page_id>/' 前缀，则可以通过 include 方法修改为：

```
urlpatterns = [
    path('<page_slug>-<page_id>/', include([
        path('history/', views.history),
        path('edit/', views.edit),
        path('discuss/', views.discuss),
        path('permissions/', views.permissions),
    ])),
]
```

7.7 向视图传递额外参数

可以使用 path() 方法的第三个参数向视图传递额外参数，例如：

```
from django.urls import path
from . import views
```

```python
urlpatterns = [
    path('blog/<int:year>/', views.year_archive, {'foo': 'bar'}),
]
```

也可以向 include() 方法传递额外参数。

配置一：

```python
# main.py
from django.urls import include, path

urlpatterns = [
    path('blog/', include('inner'), {'blog_id': 3}),
]

# inner.py
from django.urls import path
from mysite import views

urlpatterns = [
    path('archive/', views.archive),
    path('about/', views.about),
]
```

配置二：

```python
# main.py
from django.urls import include, path
from mysite import views

urlpatterns = [
    path('blog/', include('inner')),
]

# inner.py
from django.urls import path

urlpatterns = [
    path('archive/', views.archive, {'blog_id': 3}),
    path('about/', views.about, {'blog_id': 3}),
]
```

此时，额外参数 {'blog_id': 3} 将会被传递给每一个被引用的 URL。

7.8 动态生成 URL

在网页应用程序中，很多情况下需要动态编写 URL，而不是用户直接在浏览器中输入 URL，例如网页超链接的 URL 需要在生成网页时固定好。

接下来，以下 URL 模式字符串为例，看看如何在 Django 模板和视图中动态生成 URL：

```python
path('articles/<int:year>/', views.year_archive, name='news-year-archive')
```

步骤 1：使用 url 标签在模板中动态生成 URL。

```
{# 使用固定值参数：#}
```

```html
<a href="{% url 'news-year-archive' 2012 %}">2012 Archive</a>
{# 使用动态变量：#}
<ul>
{% for yearvar in year_list %}
<li><a href="{% url 'news-year-archive' yearvar %}">{{ yearvar }} Archive</a></li>
{% endfor %}
    </ul>
```

步骤 2：使用 reverse() 方法在 Python 代码中生成 URL。

```python
from django.urls import reverse
from django.http import HttpResponseRedirect

def redirect_to_year(request):
    # ...
    year = 2006
    # ...
    return HttpResponseRedirect(reverse('news-year-archive', args=(year,)))
```

7.9 URL 名字和命名空间

给 URL 命名可以方便地在模板或 Python 代码中使用 URL，如 7.8 节的示例中分别在模板和 Python 代码中使用了 URL 的名字 'news-year-archive'。

URL 命名空间用于将 URL 进行隔离。通常可以使用应用程序名作为 URL 的命名空间，例如 django.contrib.admin 的命名空间就是 admin。由于 Django 的应用程序可以部署多次，因此应用程序的实例名也可以作为命名空间。

使用 "命名空间名:URL 名" 的方式调用 URL。命名空间可以嵌套使用如 "命名空间 1: 命名空间 2:URL 名"。

步骤 1：定义命名空间。

在 URLconf 模块中使用 app_name 属性声明命名空间，例如：

```python
#!/usr/bin/python
# -*- coding: UTF-8 -*-

from django.urls import path

from . import views

app_name = 'polls'
urlpatterns = [
    ...
]
```

或者直接在 urlpatterns 中定义命名空间：

```python
from django.urls import include, path

from . import views

polls_patterns = ([
    path('', views.IndexView.as_view(), name='index'),
```

```
    path('<int:pk>/', views.DetailView.as_view(), name='detail'),
], 'polls')

urlpatterns = [
    path('polls/', include(polls_patterns)),
]
```

上面的 polls_patterns 是一个元组，元组的第一个参数是 path() 或 re_path() 列表，第二个参数是 URL 的 namespace。当使用 include() 方法引用 polls_patterns 时，系统会自动为 polls_patterns 中的所有 URL 添加 namespace。

步骤 2：在其他 URLconf 中使用命名空间。

```
from django.urls import include, path

urlpatterns = [
    path('polls/', include('polls.urls')),
]
```

步骤 3：在模板文件中使用命名空间。

```
{% url 'polls:index' %}
```

步骤 4：在 Python 代码中使用命名空间。

```
return HttpResponseRedirect(reverse('polls:results', args=(question.id,)))
```

第8章 模型

8.1 模型简介

模型是一个用于表示数据的 Python 类,包含基本的数据字段和行为,通常一个模型就代表一张数据库表。模型继承自 django.db.models.Model,模型的每一个属性代表一个数据的字段。

前面已经使用模型创建过调查问卷系统的调查问卷类和问卷选项类,接下来再用一个简单的例子介绍一下模型。下面的模型 Person 属于 myapp 应用程序:

```
from django.db import models

class Person(models.Model):
    first_name = models.CharField(max_length=30)
    last_name = models.CharField(max_length=30)
```

模型 Person 包括两个字段:first_name 和 last_name。这两个字段都是模型的类属性,分别对应数据库表中的两个列。

当执行 migrate 命令时,Django 会执行类似下面的 SQL 脚本来创建 Person 对应的数据库表:

```
CREATE TABLE myapp_person (
    "id" serial NOT NULL PRIMARY KEY,
    "first_name" varchar(30) NOT NULL,
    "last_name" varchar(30) NOT NULL
);
```

代码分析:
- Django 根据模型所属应用程序生成数据库表名,命名规则:应用程序名_模型名。
- Django 自动添加 id 字段作为数据库表的主键,与其他字段一样可以自定义主键,自定义主键需要包含 primary_Key=True,格式如下:

```
id = models.AutoField(primary_Key=True)
```

8.2 使用模型

首先在 Django 的配置文件中注册应用程序,否则 migrate 命令无法找到模型。打开 settings.py 文件,找到 INSTALLED_APPS 配置项,将模型所在应用程序名添加到列表中。一般情况下,应用程序名就是使用 startapp 命令时所填写的名字。如果忘记了应用程序名,可以到应用程序文件夹下,找到 apps.py 文件,打开即可找到应用程序名:

```python
from django.apps import AppConfig

class PollsConfig(AppConfig):
    name = '应用程序名'
```

新的配置信息如下：

```python
INSTALLED_APPS = [
    ...
    'myapp',
]
```

配置完成，执行以下命令生成对应的数据库表：

```
python manage.py makemigrations 应用程序名
python manage.py migrate
```

8.3 字段

字段是模型最重要的组成部分，它是一系列数据字段的定义。模型字段是模型的类属性，它的命名不能与模型接口相同，如不能定义名为 clean、save、delete 等，同时字段名字中不能连续出现两个下画线，这是因为连续的两个下画线是 Django 数据库 API 的特殊语法。

每一个模型字段都是 django.db.models.Field 的实例，对应一种数据库存储格式以及 HTML 元素样式。

为了支持不同的数据库，Django 提供了几十种字段类型，接下来介绍常用的几种。

8.3.1 AutoField

IntegerField 的改进形式，字段值根据已有的 id 自动增长，常用作主键。一般情况下 Django 已经帮你自动创建了主键。

8.3.2 BigAutoField

与 AutoField 相似，不过 BigAutoField 使用 64 位整型存储数据，取值范围为 1～9 223 372 036 854 775 807。

8.3.3 BinaryField

用于存储原始二进制数据的字段，如 bytes、bytearray 或者 memoryview 等 Python 类型数据。
默认情况下 BinaryField 字段是不能编辑的（editable=False），因此不能放在 ModelForm 中。
BinaryField 字段包含一个额外的可选参数 BinaryField.max_length，该参数限制了字段长度，用于表单验证。

8.3.4 BooleanField

字段值只包含 True 和 False。类似于 SQL Server 中的 bit 类型。默认情况下，BooleanField 对应 HTML 的复选框：<input type="checkbox" ...>。如果没有设置 Field.default 属性，那么它的默认值是 None。

8.3.5 CharField

字符串类型，用于保存不太长的字符串。使用该字段必须给出 CharField.max_length，该属性指定了 CharField 所能接收的最大字符数，也用于字段有效性验证。默认情况下，CharField 对应 HTML 的文本框：<input type="text" ...>。

> 由于不同数据库对字符串字段的大小限制不一样，因此在设置 max_length 的时候要考虑自己的数据库特性。

对于超长的字符串，建议使用 TextField 类型。

8.3.6 DateField

日期类型，对应 Python 中的 datetime.date 类型。该字段类型包含以下几个可选参数。

- DateField.auto_now：如果指定了该属性，那么每当保存数据时都会将该字段值更新为当前时间，auto_now 与默认值相似，但是又不相同。只有在调用 Model.Save() 方法的时候该字段才会被自动更新，使用其他方法更新其他字段的时候 auto_now 字段不会被更新。例如，调用 QuerySet.update() 方法的时候不会自动更新 auto_now 字段，不过此时仍然可以通过赋值更新它。
- DateField.auto_now_add：与 auto_now 类似，不过只有当该行数据被第一次创建时才会保存当前时间，主要用于记录对象的创建时间。不像默认值，标记为 auto_now_add 的字段值是不能被重写的，如果确实需要一个既能使用当前时间又能被编辑的字段，可以按照如下方式设置：

```
DateField: default=date.today
DateTimeField: default=timezone.now
```

如果将日期类型字段的 auto_now 或者 auto_now_add 属性设置为 True，那么字段的 editable 属性会被自动设置为 False，同时 blank 属性会被自动设置为 True。

auto_now、auto_now_add 和 default 三个参数只能单独存在，不能搭配设置，否则会引起异常。

auto_now、auto_now_add 属性总是使用 TIME_ZONE 中设置的时区保存时间，如果想使用其他时区，则不应该设置这两个属性。

示例：

```
date_field = models.DateField(default=datetime.datetime.now())
```

显示效果如下图所示。

单击日历图表后，弹出日历控件，效果如下图所示。

8.3.7 DateTimeField

日期时间类型，对应 Python 中的 datetime.datetime 类型。与 DateField 一样，包含两个额外参数 auto_now 和 auto_now_add。

默认情况下，DateTimeField 对应 HTML 的文本框：<input type="text" ...>，在 Admin 后台页面则使用两个文本框显示 DateTimeField 字段。

示例：

```
date_time_field = models.DateTimeField(default=datetime.datetime.now())
```

显示效果如下图所示。

单击钟表图表后，弹出预设时间选择控件，效果如下图所示。

8.3.8 DecimalField

指定小数位数的数值类型，对应 Python 的 Decimal 类型。
该字段类型包含两个必要参数：
- DecimalField.max_digits，数值的总位数，例如 123.45 就是 5 位。
- DecimalField.decimal_places，小数点后位数。

例如，设置一个字段的最大值为 999，同时包含两位小数：

```
models.DecimalField(max_digits=5, decimal_places=2)
```

DecimalField 对应 HTML 的文本框：<input type= "number" ...>，如果禁用了本地化，那么 DecimalField 对应 HTML 的控件是 <input type= "text" ...>。

示例：

```
decimal_field = models.DecimalField(max_digits=5, decimal_places=2, default=0.00)
```

显示效果如下图所示。

8.3.9　EmailField

本质就是 CharField，不过会验证输入的字符串是不是一个有效的邮件地址。

示例：

```
email_field = models.EmailField(default='test@test.com')
```

显示效果如下图所示。

8.3.10　FileField

文件上传控件。

该字段不允许使用 primary_Key 属性。包含两个可选参数：FileField.upload_to 和 FileField.storage。

FileField.upload_to 表示文件上传后的保存位置，例如：

```
class MyModel(models.Model):
        # 文件上传到 MEDIA_ROOT/uploads
        Upload1 = models.FileField(upload_to='uploads/')
        # 文件上传到 MEDIA_ROOT/uploads/2017/12/30
        Upload2 = models.FileField(upload_to='uploads/%Y/%m/%d/')
```

> **注意**
>
> MEDIA_ROOT 在 settings.py 中设置，upload_to 所指定的路径将会拼接在 MEDIA_ROOT 之后。
>
> 有时需要通过可编程的方式决定文件的保存路径，可以给 upload_to 指定一个可调用对象，该对象接收两个参数：
> ❑ Instance：包含文件的模型实例。
> ❑ Filename：文件的初始名字。

示例：

```
def user_directory_path(instance, filename):
    # file will be uploaded to MEDIA_ROOT/user_<id>/<filename>
```

```
        return 'user_{0}/{1}'.format(instance.user.id, filename)

class MyModel(models.Model):
    upload = models.FileField(upload_to=user_directory_path)
```

FileField.storage 是负责文件存储的 Python 类，用于存储和提取文件。

django.core.files.storage.FileSystemStorage 类提供了基本的文件管理功能。例如下面的代码，无论在 settings.py 里面如何设置 MEDIA_ROOT 都会按照 fs 所指定的路径存储被上传的文件：

```
from django.db import models
from django.core.files.storage import FileSystemStorage

fs = FileSystemStorage(location='/media/documents')

class Car(models.Model):
    ...
    photo = models.FileField(storage=fs)
```

FileField 对应 HTML 的文件上传控件：<input type="file" ...>。

示例：

```
file_Field = models.FileField(upload_to='file_field')
```

显示效果如下图所示。

| File Field: | 选择文件 未选择任何文件 |

8.3.11 FilePathField

文件列表显示字段，列表内容必须是某一个路径下的文件。

参数介绍如下：

- FilePathField.path，必选参数，文件夹的绝对路径，该路径下的文件将会显示在下拉列表中。Django 3.0 中 path 参数可接收一个可调用对象。
- FilePathField.match，可选参数，一段正则表达式，用于过滤 FilePathField.path 中的文件，注意该正则表达式只能过滤文件名，不能过滤路径。
- FilePathField.recursive，可选参数，参数值为 True 或 False。当参数设置为 True 时，FilePathField.path 的子文件夹中符合条件的文件也会显示在下拉列表中，否则只显示当前文件夹中的文件。默认值为 True。
- FilePathField.allow_files，可选参数，规定是否包含文件，参数值为 True 或 False，默认值为 True，allow_files 和 allow_folders 之间必须有一个为 True。
- FilePathField.allow_folders，可选参数，规定是否包含文件夹，参数值为 True 或 False，默认值为 True，allow_files 和 allow_folders 之间必须有一个为 True。

接下来通过下面的目录结构查看 FilePathField 显示效果。

```
File/
    file_path/
        css.jpg
        linux.jpg
```

```
                    Django/
                           Django.jpg
```

示例1：

```
file_path_field = models.FilePathField(path='D:/File/file_path', match='.jpg$',
recursive=False, allow_files=True, allow_folders=False)
```

显示效果如下图所示。

示例2：

```
file_path_field = models.FilePathField(path='D:/File/file_path', match='.jpg$',
recursive=True, allow_files=True, allow_folders=False)
```

显示效果如下图所示。

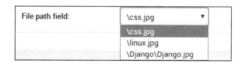

> **注意**
>
> 文件的绝对路径会显示在 HTML 代码中，这会产生一定的安全隐患，因此需要谨慎使用 FilePathField 对象。
>
> FilePathField 在数据库中使用 varchar 类型存储，默认最大长度为 100 字符。

8.3.12　FloatField

浮点数类型，对应 Python 的 float 类型。

8.3.13　ImageField

继承 FileField 字段的全部属性与方法，但是仅允许上传图片类型文件。

为了设置图片显示的高度与宽度，ImageField 字段额外提供了两个属性：
- ImageField.height_field，图片高度。
- ImageField.width_field，图片宽度。

8.3.14　IntegerField

整数字段，取值范围为 -2 147 483 648 ～ 2 147 483 647。该类型对于所有 Django 支持的数据库来说都是安全的。

8.3.15　GenericIPAddressField

IPv4 或者 IPv6 字段类型（例如 192.0.2.30 或者 2a02:42fe::4）。对应的 HTML 控件是 <input

type="text" ...>。

8.3.16 PositiveIntegerField

正整数类型，取值范围为 0 ～ 2 147 483 647。

8.3.17 PositiveSmallIntegerField

小正整数类型，取值范围为 0 ～ 32 767。

8.3.18 SlugField

Slug 是新闻业的专业名词，代表一个简短的文本，只允许包含字母、数字、下画线和连字符。与 CharField 相似，可以指定 max_length 属性，如果没有显式给出 max_length 值，默认值为 50。

如果想在 SlugField 字段中使用除 ASCII 之外的其他 Unicode 字符，可以将属性 SlugField.allow_unicode 设置为 True。

8.3.19 SmallIntegerField

小整数类型，取值范围为 -32 768 ～ 32 767。

8.3.20 TextField

超长文本类型。

示例：

```
text_field = models.TextField(default='')
```

显示效果如下图所示。

8.3.21 TimeField

时间类型，对应 datetime.time。

8.3.22 URLField

CharField 类型，只能接受 URL 字符串，默认最大长度是 200 个字符。

8.3.23 UUIDField

UUID 类型字段，对应 Python 的 UUID 类。

现代数据库应用中更推荐使用 uuid 作为主键而不是自增字段。

示例：

```
import uuid
from django.db import models

class MyUUIDModel(models.Model):
    id = models.UUIDField(primary_key=True, default=uuid.uuid4, editable=False)
```

8.4 字段参数

每一个字段都需要一系列参数，例如，使用 CharField 时必须给出 max_length 参数，除了以上特殊字段参数外，Django 还为所有字段提供了一系列通用参数，这些参数都是可选的。

接下来详解字段的通用参数。

8.4.1 null

如果设置为 True，当该字段为空时，Django 会将数据库中该字段设置为 NULL。默认为 False。

对于文本型字段，尽可能不使用 null 属性，因为当使用默认值 null 时，数据库中就可能出现两种空数据：NULL 和空字符串，而 Django 默认使用空字符串。

8.4.2 blank

默认值为 False，当设置 Field.blank=True 时字段值允许为空。

与 Field.null 属性不同的是，null 只是表示数据库值是空的，而 blank 用于表单验证。当字段属性 blank=True 时，表单验证将允许字段值为空，但是，当 blank=False 时，表单字段将变成必填字段。

8.4.3 choices

属性值为一个可迭代对象，如列表或者元组，迭代对象的每个成员包括两个元素。当字段设置 choices 属性时，字段在网页中将会以下拉列表的形式显示。

列表或元组的第一个值将作为字段值保存到数据库中，第二个值用于提高字段的可读性。

示例 1：

```
YEAR_IN_SCHOOL_CHOICES = [
    ('FR', 'Freshman'),
    ('SO', 'Sophomore'),
    ('JR', 'Junior'),
    ('SR', 'Senior'),
    ('GR', 'Graduate'),
]
year_in_school = models.CharField(
        max_length=2,
        choices=YEAR_IN_SCHOOL_CHOICES,
        default='Freshman',
        )
```

显示效果如下图所示。

示例2:

```
MEDIA_CHOICES = [
    ('Audio', (
            ('vinyl', 'Vinyl'),
            ('cd', 'CD'),
        )
    ),
    ('Video', (
            ('vhs', 'VHS Tape'),
            ('dvd', 'DVD'),
        )
    ),
    ('unknown', 'Unknown'),
]
group_choice_field = models.CharField(
        max_length=10,
        choices=MEDIA_CHOICES,
        default='Audio',
    )
```

显示效果如下图所示。

示例3:

修改示例2,设置字段 blank 属性为 True:

```
group_choice_field = models.CharField(
        max_length=10,
        choices=MEDIA_CHOICES,
        default='Audio',
        blank=True,
    )
```

显示效果如下图所示。

示例 4：

修改示例 2，在元素 MEDIA_CHOICES 中添加成员 (None, 'Please Select Media')：

```
MEDIA_CHOICES = (
    (None, 'Please Select Media'),
    ('Audio', (
            ('vinyl', 'Vinyl'),
            ('cd', 'CD'),
        )
    ),
    ('Video', (
            ('vhs', 'VHS Tape'),
            ('dvd', 'DVD'),
        )
    ),
    ('unknown', 'Unknown'),
)
```

显示效果如下图所示。

Django 3.0 中新增了 TextChoices、IntegerChoices 和 Choices 三个类型，以 TextChoices 为例演示使用效果。

编写代码：

```
from django.utils.translation import gettext_lazy as _

class Student(models.Model):

    class YearInSchool(models.TextChoices):
        FRESHMAN = 'FR', _('Freshman')
        SOPHOMORE = 'SO', _('Sophomore')
        JUNIOR = 'JR', _('Junior')
        SENIOR = 'SR', _('Senior')
        GRADUATE = 'GR', _('Graduate')
```

```
        year_in_school = models.CharField(
            max_length=2,
            choices=YearInSchool.choices,
            default=YearInSchool.FRESHMAN,
        )
```

显示效果如下图所示。

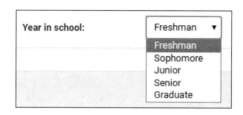

8.4.4 default

该字段的默认值，可以是一个值或者一个可调用对象，如果是一个可调用对象，每次实例化模型时都会调用该对象。

8.4.5 help_text

额外的"帮助"文本，随表单控件一同显示，在文本中可以使用 HTML 标记。
示例：

```
help_text = "Please select your <em>favorite</em> media"
```

显示效果如下图所示。

8.4.6 primary_Key

如果设置为 True，将该字段设置为模型的主键。如果模型中任何字段都不包含 primary_Key=True 属性，Django 将会自动为模型添加一个 IntegerField 字段作为主键。
主键永远是只读的，如果修改一个模型实例的主键并保存，这等同于创建了一个新的模型实例。

8.4.7 unique

当字段的 unique 属性设置为 True 时，该字段的所有值在整张数据表中不能重复，每一行数据都必须有唯一的字段值。

8.4.8 verbose_name

verbose_name 属性类似于字段的说明。
除了 ForeignKey、ManyToManyField、OneToOneField 三种字段类型外，其他字段类型都包含一个默认的 verbose_name 属性，可以直接在字段属性列表的第一位输入文本作为 verbose_name 属性值。

如果没有给出 verbose_name 属性，Django 会使用字段名作为 verbose_name 值，如果字段名中包含下画线，下画线会被转换为空格。

ForeignKey、ManyToManyField、OneToOneField 三种字段类型要求第一个参数必须是模型类，因此必须使用 verbose_name 关键字。

示例：

```
date_field = models.DateField('This is publish day', default=datetime.datetime.now())
Key = models.ForeignKey(Question, verbose_name='Foreign Key')
```

8.5 表与表之间关系

关系数据库的表与表之间往往存在一定的关系，由于 Django 的模型是数据库表与 Python 类之间的映射，因此 Django 提供了对三种最常用的数据库表之间关系的支持：多对一、多对多、一对一。

8.5.1 多对一关系

多对一关系是一张数据表中的多条记录与另一张数据表中的一条记录相关的关联模式。在关系数据库中通常使用外键来表示多对一关系，Django 模型中的 ForeignKey 字段就是模型的外键。

与其他模型字段的使用方法基本相同，唯一不同之处是 ForeignKey 类的第一个参数是关联的类名。例如，前面示例代码中 Choice 类的外键 question 的第一个参数是 Question 类名：

```
class Choice(models.Model):
    question = models.ForeignKey(Question, on_delete=models.CASCADE)
    ...
```

在调查问卷系统中，每一个调查问卷都可以包含多个问卷选项，而每一个选项只能属于一个调查问卷，因此选项与问卷之间形成了多对一的关系。

在数据库中查看 polls_choice 表结构，如下图所示。

名	类型	长度	小数点	不是 null	
id	int	11	0	✓	🔑 1
choice_text	varchar	200	0	✓	
votes	int	11	0	✓	
▶ question_id	int	11	0	✓	

虽然在模型 Choice 类中并没有定义 question_id 字段，但是 Django 自动创建了 question_id 字段作为 Choice 的外键。

8.5.2 多对多关系

另一种比较常见的数据库表之间的关系是多对多，如前面示例中调查问卷与问卷分组之间的关系就是多对多。如问卷可以按照创建时间进行分组，也可以按照问卷类型进行分组。某一个问卷既可以属于"2020 年"组，又可以属于"IT 类"分组。

Django 自带的 Auth 模块中存在大量多对多的数据关系，例如用户与用户组之间、用户与用户权限之间、用户组与用户权限之间。对于多对多关系，Django 会在数据库中额外创建一张关系表，关系表的命名规则是：应用程序名 _ 模型 1 名 _ 模型 2 名，例如用户与用户组的关系表就叫作 auth_user_groups。数据库中查看 auth_user_groups 表结构，如下图所示。

在 Django 中使用多对多关系时，有以下几个建议：
- 多对多字段名使用复数形式。
- 可以在两个有多对多关系的模型中的任意一个模型中定义多对多字段，但是不能同时在两个模型中都定义多对多字段。

8.5.3 一对一关系

这种映射关系用得比较少，Django 使用 OneToOneField 表示一对一关系。

一对一关系的一个比较常用的场景是根据一张表的主键对这张表进行扩展，例如对 Django 自带的 user 表进行扩展，为每一个用户数据添加额外信息。

与前两种关系字段的使用相同，OneToOneField 也要求第一个参数是模型类名。

8.6 模型元属性

元属性是"模型中任意非模型字段的内容"，例如排序功能、数据表名、人类可读的名字（单数形式和复数形式），所有元属性都是可选的。

通过在模型中添加一个叫作 Meta 的子类，定义模型元属性。

在详细介绍模型元属性之前，先看一下如何在 Django 模型中使用元属性：

```
from django.db import models

class Ox(models.Model):
    horn_length = models.IntegerField()

    class Meta:
        ordering = ["horn_length"]
        verbose_name_plural = "oxen"
```

通过设置元属性，上面代码中的 Ox 模型将会默认使用 horn_length 排序，在 Admin 页面显示的复数形式名字叫 oxen。

8.7 元属性

模型的元数据即"所有不是字段的东西"，例如排序选项（ordering）、数据库表名（db_table）或者阅读友好的单复数名（verbose_name 和 verbose_name_plural）。这些都不是必需的，并且在模型当中添加 Meta 类也是可选的。下面介绍部分常用模型元属性。

8.7.1 abstract

如果设置 abstract = True，那么当前模型将成为一个抽象类。例如：

```
from django.db import models

class CommonInfo(models.Model):
    name = models.CharField(max_length=100)
    age = models.PositiveIntegerField()

    class Meta:
        abstract = True

class Student(CommonInfo):
    home_group = models.CharField(max_length=5)
```

8.7.2 app_label

如果模型所在位置不属于任何已注册的应用程序，即模型不在 INSTALLED_APPS 范围内，那么必须使用 app_label 选项来指定所属的应用程序名字。

8.7.3 base_manager_name

Django 访问关联对象时所使用的管理器实例名，例如 Choice.objects.filter(question__name__startswith='Django') 中的 objects。

8.7.4 db_table

当前模型所使用的数据表名。

默认情况下，Django 会自动地根据应用程序名 + 模型名生成数据库表名。例如前面示例代码中的 polls 应用程序，生成的数据表叫作 polls_question、polls_choice。

如果你觉得 Django 自动生成的表名不好看，就可以通过 db_table 来重新定义表名。在这里即使表名不合法也没关系，Django 会替你处理它，不过最好还是使用规规矩矩的名字。

8.7.5 get_latest_by

Options.get_latest_by 是一组字段名，Django 的模型管理器会根据这些字段查找最新的数据，get_latest_by 通常使用 DateField、DateTimeField 或者 IntegerField 这几种字段类型。如果在字段前添加减号，表示降序排列，例如按照 priority 降序排列，按照 order_date 升序排列：

```
get_latest_by = ['-priority', 'order_date']
```

8.7.6 order_with_respect_to

使对象能够根据某一个字段进行排序，通常是外键。例如可以使问卷选项根据调查问卷进行排序。

修改 Choice 模型类，增加 order_with_respect_to 属性：

```
class Choice(models.Model):
    ...
    class Meta:
        order_with_respect_to = 'question'
```

执行 migrate 命令使得变更应用到数据库。

设置完 order_with_respect_to 后，Choice 的相关对象（related object，即 Question）将获得两个新方法：get_RELATED_order() 和 set_RELATED_order()，本例中方法名中间的 RELATED 对应小写的 Question。Choice 本身也获得了两个新方法：get_next_in_order() 和 get_previous_in_order()。

在交互窗口中测试以上新方法，如下图所示。

> **注意**
>
> 当问卷选项已经是序列中的第一位时调用 get_previous_in_order() 方法或者已经是序列中最后一位时调用 get_next_in_order() 方法将出现错误：
>
> ```
> polls.models.Choice.DoesNotExist: Choice matching query does not exist.
> ```

8.7.7　ordering

默认的排序字段，当从数据库中查找数据时会按照 ordering 指定的排序方式显示。

该属性可以是元组、列表或者查询表达式，每一个元组或列表的元素都是一个字段名，默认是正序排序，如果给字段名前添加一个"-"符号则会按照倒序排序；如果在字段名前面加"?"，就会随机提取数据。

例如，按照 pub_date 倒序、author 正序查找数据：

```
ordering = ['-pub_date', 'author']
```

使用查询表达式根据 author 字段进行排序，当 author 字段值为 null 时，数据放在最后显示：

```
from django.db.models import F

ordering = [F('author').asc(nulls_last=True)]
```

8.7.8　Indexes

用来定义数据库索引，形式如下：

```python
from django.db import models

class Customer(models.Model):
    first_name = models.CharField(max_length=100)
    last_name = models.CharField(max_length=100)

    class Meta:
        indexes = [
            models.Index(fields=['last_name', 'first_name']),
            models.Index(fields=['first_name'], name='first_name_idx'),
        ]
```

8.7.9 constraints

一组模型字段的约束条件，例如以下代码可以限制客户年龄大于或等于 18 岁：

```python
from django.db import models

class Customer(models.Model):
    age = models.IntegerField()

    class Meta:
        constraints = [
            models.CheckConstraint(check=models.Q(age__gte=18), name='age_gte_18'),
        ]
```

8.7.10 verbose_name

便于人类读取的模型名称，单数形式。

代码形式如下：

```
verbose_name = "pizza"
```

如果没有指定 verbose_name，Django 会根据模型类名自动创建 verbose_name，如模型 CamelCase 对应的 verbose_name 为 camel case。

8.7.11 verbose_name_plural

便于人类读取的模型名称，复数形式。

代码形式如下：

```
verbose_name_plural = "stories"
```

如果没有指定 verbose_name_plural，Django 会自动生成，形式为 verbose_name + "s"。

8.8 Manager 类

Manager 是 Django 模型最重要的属性，它赋予模型操作数据库的能力。默认情况下，Django 会为每一个模型提供一个叫作 objects 的 Manager 实例。Manager 属性只能通过模型类访问。

8.8.1 自定义 Manager 类

默认情况下可以使用 Model.objects 所提供的方法操作数据库,也可以实现自定义 Manager 类:

```
from django.db import models

class Person(models.Model):
    #...
    people = models.Manager()
```

此时查找数据的方式如下:

```
Person.people.all()
```

8.8.2 直接执行 SQL 语句

虽然 Manager 类非常强大,但是有些情况下,仍然希望手写 SQL 语句。对此 Django 提供了两种方法允许开发人员直接执行 SQL 语句。

方法 1:使用 Manager.raw() 方法。

Manager.raw() 方法可以用来执行一段 SQL 语句并返回 Django 模型实例。

语法:Manager.raw(raw_query, params=None, translations=None)

raw() 方法返回一个 django.db.models.query.RawQuerySet 实例,RawQuerySet 与前面的 QuerySet 一样可以被循环遍历。

下图是一个在 shell 中使用 raw() 方法查看全部调查问卷的例子。

```
PS D:\Code\django3\mysite> python .\manage.py shell
Python 3.8.0 (tags/v3.8.0:fa919fd, Oct 14 2019, 19:21:23) [MSC v.1916 32 bit (Intel)] on win32
Type "help", "copyright", "credits" or "license" for more information.
(InteractiveConsole)
>>> from polls.models import *
>>> questions = Question.objects.raw("select * from polls_question")
>>> for q in questions: print(q.question_text)
...
你喜欢Django吗?
你学习Django多久了?
>>>
```

Django 不会对 raw() 方法中执行的 SQL 语句进行检查,但是 Django 希望 SQL 语句能够返回一行或多行数据,如果执行结束没有返回任何数据,raw() 方法将会抛出异常。

Manager.raw() 方法能够自动将查询结果转换为对应的模型,即使在查询语句中使用 AS 关键字对字段名进行了修改也没关系,只要数据库中的字段名与模型字段匹配成功即可,效果如下图所示。

```
>>> for q in questions: print(type(q))
...
<class 'polls.models.Question'>
<class 'polls.models.Question'>
```

由于 Manager.raw() 方法返回 RawQuerySet 对象,因此可以通过索引的方式遍历结果,效果如下图所示。

```
>>> questions[0]
<Question: 你喜欢Django吗?>
```

如果需要提取 SQL 语句中不包含的字段时，Manager.raw() 将对被省略的字段进行按需加载，查询返回的对象即延迟模型实例，效果如下图所示。

```
>>> questions = Question.objects.raw("select id,question_text from polls_question")
>>> questions[1].pub_date
datetime.datetime(2020, 2, 18, 8, 49, tzinfo=<UTC>)
>>> questions[1].description
'你学习Django多久了？'
```

以上例子虽然看起来能够正确查询所有字段，但是实际上 Django 进行了 3 次查询：
（1）执行 raw 命令查询 id 和 question_text。
（2）根据步骤 1 得到的 id 查询 pub_date。
（3）根据步骤 1 得到的 id 查询 description。

看到这里你可能就明白了，如果需要进行延迟查询，那么在 raw 方法中必须包含主键，另外延迟查询会增加额外的数据库交互。是否需要进行延迟查询还要看你的业务要求。

所执行的 SQL 语句中除了可以包含模型字段外，还可以包含其他聚合函数值，如在查找调查问卷的同时输出问卷标题所包含的字符数（length 是 MySQL 和 SQLite 等数据库提供的函数），如下图所示。

```
>>> sql = "select id,question_text,length(question_text) as l from polls_question"
>>> questions = Question.objects.raw(sql)
>>> print(questions[0].question_text, questions[0].l)
你喜欢Django吗？ 11
```

Manager.raw() 方法还可以对 SQL 语句进行参数化，参数可以是列表或者字典，如下图所示。

```
>>> sql = "select * from polls_question where id=%s"
>>> questions = Question.objects.raw(sql, [1])
>>> print(questions[0].question_text)
你喜欢Django吗？
```

或者如下图所示。

```
>>> arg = {'d':1}
>>> sql = "select * from polls_question"
>>> questions = Question.objects.raw(sql, arg)
>>> questions[0].question_text
'你喜欢Django吗？'
```

为了避免 SQL 注入攻击，不应该使用字符串格式化工具对 SQL 语句进行编辑，应当统一使用 raw 方法的 params 参数接收变量。以下形式都可能为系统带来安全隐患：

```
query1 = 'SELECT * FROM myapp_person WHERE last_name = %s' % lname
Person.objects.raw(query1)
query2 = "SELECT * FROM myapp_person WHERE last_name = '%s'" % lname
Person.objects.raw(query2)
```

方法 2：脱离模型，直接执行 SQL。

由于 Manager.raw() 的执行结果总是对应一个模型，而真正软件产品中不只是查询单个模型，还会有更复杂的情况，例如执行更新、删除、插入等操作。因此需要跳出模型系统直接执行 SQL 语句。

django.db.connection 对象提供了数据库连接操作，使用 connection.cursor() 方法可以得到一个游标对象，cursor.execute(sql, [params]) 方法用于执行指定的 SQL 语句。使用 cursor.fetchone() 或者 cursor.fetchall() 方法可以得到一个或全部结果。

示例：

```
from django.db import connection

def my_custom_sql(self):
    with connection.cursor() as cursor:
        cursor.execute("UPDATE bar SET foo = 1 WHERE baz = %s", [self.baz])
        cursor.execute("SELECT foo FROM bar WHERE baz = %s", [self.baz])
        row = cursor.fetchone()

    return row
```

上述代码中 SQL 语句内变量 %s 没有使用单引号包围起来，这样做对防止 SQL 注入有一定的作用。

当 SQL 语句需要传递参数并且确实需要包含一个百分号时，那么可以输入两个百分号，例如：

```
cursor.execute("SELECT foo FROM bar WHERE baz = '30%'")
cursor.execute("SELECT foo FROM bar WHERE baz = '30%%' AND id = %s", [self.id])
```

如果当前工程包含多个数据库，可以使用 django.db.connections 对象获取特定数据库的连接。django.db.connections 是一个类似字段对象，字典的 Key 是数据库的别名，例如在 setting 中指定 polls 所使用的数据库别名是"polls"，那么连接 polls 可以使用 connections['polls']。

```
DATABASES = {
    'polls': {
        'ENGINE': 'django.db.backends.sqlite3',
        'NAME': os.path.join(BASE_DIR, 'db.sqlite3'),
    }
}
```

代码示例如下图所示。

```
>>> from django.db import connections
>>> connections.databases.keys()
dict_keys(['default', 'polls'])
```

需要注意的是，cursor 所返回的数据只是数据库中所有字段的值，也就是一个数值的列表，而不是一个同时包含字段名与字段值的字典，为了使返回的数据方便使用，可以使用下面方法将返回结果转换为字典：

```
def dictfetchall(cursor):
    "Return all rows from a cursor as a dict"
    columns = [col[0] for col in cursor.description]
    return [
        dict(zip(columns, row))
        for row in cursor.fetchall()
    ]
```

使用 cursor 重新在 shell 中执行查询操作，效果如下图所示。

```
>>> from django.db import connections
>>>
>>> def dictfetchall(cursor):
...     "Return all rows from a cursor as a dict"
...     columns = [col[0] for col in cursor.description]
...     return [
...         dict(zip(columns, row))
...         for row in cursor.fetchall()
...     ]
...
>>>
>>> with connections['polls'].cursor() as cursor:
...     cursor.execute("SELECT * FROM polls_question")
...     data = dictfetchall(cursor)
...
<django.db.backends.sqlite3.base.SQLiteCursorWrapper object at 0x044A3528>
>>>
>>> print(data)
[{'id': 1, 'question_text': '你喜欢Django吗？', 'pub_date': datetime.datetime(2020, 1, 11, 9
': 5, 'question_text': '你学习Django多久了？', 'pub_date': datetime.datetime(2020, 2, 18, 8,
>>>
```

8.8.3 执行存储过程

语法：CursorWrapper.callproc(procname, params=None, kparams=None)

注意，只有 Oracle 支持 kparams 参数。

假设存储过程如下：

```
CREATE PROCEDURE "TEST_PROCEDURE"(v_i INTEGER, v_text NVARCHAR2(10)) AS
    p_i INTEGER;
    p_text NVARCHAR2(10);
BEGIN
    p_i := v_i;
    p_text := v_text;
    ...
END;
```

调用存储过程的示例代码如下：

```
with connection.cursor() as cursor:
    cursor.callproc('test_procedure', [1, 'test'])
```

8.9 数据增删改查

软件系统的核心就是对数据的增删改查，Django 通过模型以及 QuerySet API 为用户提供了丰富的数据库操作方法。

当创建好模型后，就可以立即进行添加、删除、更新、查找操作。下面通过一个例子展示如何直接使用 Django 模型类进行数据操作。

继续使用 MySite 工程，在工程下创建一个 blog 应用程序：

```
python manage.py startapp blog
```

首先创建一个 Blog 模型：

```
from django.db import models

class Blog(models.Model):
```

```python
    name = models.CharField("标题", max_length=100)
    body = models.TextField("内容")

    def __str__(self):
        return self.name
```

接下来创建一个用于添加博客文章的视图,每次访问视图的时候都会创建一篇新的博客文章,并且将新的博客文章传给模板显示。

```python
from django.http import HttpResponse
from django.template.loader import render_to_string
from .models import Blog

def create_blog(request, title, body):
    b = Blog(name=title, body=body)
    b.save()

    context = {
        "blog": b
    }
    rendered = render_to_string("create.html", context)
    return HttpResponse(rendered)
```

由于这里的博客应用程序将会使用多个页面,为了减少代码重复,首先创建一个基类模板 blog/templayes/base.html:

```html
{% load static %}
<!DOCTYPE HTML>
<html>
<head>
    <meta http-equiv="Content-Type" content="text/html; charset=gb2312" />
    <meta http-equiv="Content-Language" content="zh-cn" />
    <title>{% block title %}My amazing site{% endblock %}</title>
    <link rel="stylesheet" type="text/css" href="{% static 'css/myblog.css' %}" />
</head>

<body>
    <div id="content">
        {% block content %}{% endblock %}
    </div>
</body>
</html>
```

添加模板 blog/templayes/create.html,在模板中显示文章内容:

```html
{% extends "base.html" %}
{% block content %}

<h2 class="blog_head">{{ blog.id }} - {{ blog.name }}</h2>
<p class="blog_body">
    {{ blog.body }}
</p>

{% endblock %}
```

添加 CSS 样式文件 blog/static/css/myblog.css：

```css
.blog_head {
    background-color: rgb(45, 243, 243);
    border: 1px black solid;
    width: 400px;
    margin-bottom: 3px;
}
.blog_body {
    color: rgb(13, 27, 230);
    background-color: rgb(157, 158, 158);
    border: 1px black solid;
    min-height: 50px;
    width: 400px;
    margin-bottom: 10px;
}
```

添加 URL，blog/urls.py：

```python
#!/usr/bin/python
# -*- coding: UTF-8 -*-

from django.urls import path

from . import views
app_name = 'blog'

urlpatterns = [
    path(r'create/<str:title>/<str:body>/', views.create_blog, name='create'),
]
```

将 blog/urls.py 引入 mysite/urls.py：

```python
from django.contrib import admin
from django.urls import include, path

urlpatterns = [
    path('polls/', include('polls.urls')),
    path('blog/', include('blog.urls')),
    path('admin/', admin.site.urls),
]
```

在 settings.py 中注册 blog 应用程序：

```python
INSTALLED_APPS = [
    ...
    'blog.apps.BlogConfig',
]
```

在 settings.py 中配置静态文件：

```python
STATIC_URL = '/static/'
```

生成数据库：

```
python manage.py makemigrations blog
python manage.py migrate blog
```

启动网站：

```
python manage.py runserver
```

在浏览器中访问该视图：
http://127.0.0.1:8000/blog/create/Django 介绍 /Django 是基于 Python 开发的 Web 框架
显示如下图所示。

由上可见，非常简单的几行代码就完成了数据添加操作。那么数据是不是真的写入数据库了呢？打开数据库，可以看到确实完成了创建新博客的工作，如下图所示。

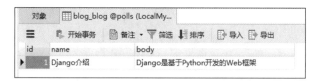

下面使用 Django objects 接口分别实现查找指定文章以及查找全部文章的视图。
在 blog/views.py 中创建 index 视图用于显示全部文章：

```
def index(request):
    blogs = Blog.objects.all()

    context = {
        "blogs": blogs
    }

    rendered = render_to_string("index.html", context)
    return HttpResponse(rendered)
```

创建 blog/templayes/index.html 模板：

```
{% extends "base.html" %}
{% load static %}

{% block content %}

    {% for blog in blogs %}
        <h2 class="blog_head">{{ blog.id }} - {{ blog.name }}</h2>
        <p class="blog_body">
            {{ blog.body }}
        </p>
    {% endfor %}

{% endblock %}

<p>{{ blog.body }}</p>
```

为 blog/urls.py 添加 url：

```
path('', views.index, name='default'),
path(r'index/', views.index, name='index'),
```

打开浏览器查看视图，效果如下图所示。

上面的视图会显示全部博客信息，实际工作中往往还需要显示某一篇文章的详情而不是使用列表显示文章的概述，下面创建一个 search_blog 视图：

```
def search_blog(request, blog_id):
    blog = Blog.objects.filter(id=blog_id)

    context = {
        "blog": blog[0]
    }

    rendered = render_to_string("detail.html", context)
    return HttpResponse(rendered)
```

创建 blog/templayes/detail.html 模板：

```
{% extends "base.html" %}

{% block title %}My amazing blog{% endblock %}

{% block content %}
<div class="center">
    <h2 class="blog_head">{{ blog.id }} - {{ blog.name }}</h2>
    <p class="blog_body">
        {{ blog.body }}
    </p>
</div>
{% endblock %}
```

添加 url：

```
path(r'<int:blog_id>/', views.search_blog, name='detail'),
```

在浏览器里面查看 id 为 1 的博客文章，效果如下图所示。

下面编写 update_blog 视图用于更新指定文章：

```
def update_blog(request, id, title='', body=''):
    blog = Blog.objects.get(id=id)
    blog.body = body
    blog.title = title
    blog.save()

    blog = Blog.objects.get(id=id)

    context = {
        "blog": blog
    }

    rendered = render_to_string("detail.html", context)
    return HttpResponse(rendered)
```

添加 url：

```
path(r'update/<int:id>/<str:title>/<str:body>/', views.update_blog, name='update'),
```

在浏览器中访问 update_blog 视图并更新 id 为 1 的文章，url 地址如下：

http://127.0.0.1:8000/blog/update/1/Django 介绍 /Django 是一套开放源代码的 Web 应用框架，由 Python 写成。采用了 MTV 的框架模式，即模型 M、视图 V 和模板 T。它最初是被开发用于管理劳伦斯出版集团旗下的一些以新闻内容为主的网站的，即 CMS（内容管理系统）软件。并于 2005 年 7 月在 BSD 许可证下发布。这套框架是以比利时的吉普赛爵士吉他手 Django Reinhardt 来命名的。

输入以上这个超长的 URL，结果如下图所示。

最后来看删除操作，同样创建一个视图：

```
def delete_blog(request, id):
    Blog.objects.get(id=id).delete()

    blogs = Blog.objects.all()

    context = {
        "blogs": blogs
    }

    rendered = render_to_string("index.html", context)
    return HttpResponse(rendered)
```

添加 url：

```
path(r'delete/<int:id>/', views.delete_blog, name='delete'),
```

首先通过 create_blog 视图添加一篇测试文章，如下图所示。

在浏览器中访问 delete_blog 视图并删除 id 为 2 的文章，操作结束返回首页显示剩余的全部文章，如下图所示。

8.10 数据操作进阶——QuerySets

8.9 节使用模型的 objects 属性完成数据的增删改查操作，如 save()、all()、filter()、get()、delete()，这些方法是 QuerySet 对象所提供的常用方法。模型类的 Manage 对象可以创建 QuerySet 实例，默认的 Manage 对象实例名为 objects。QuerySet 是数据库中一系列数据的集合，与 SQL 查询语句一样，QuerySet 可以接收 0 个、1 个或多个过滤条件。

本节将使用以下示例代码对 QuerySet 进行详细介绍：

```python
from django.db import models

class Blog(models.Model):
    name = models.CharField(max_length=100)
    tagline = models.TextField()

    def __str__(self):
        return self.name

class Author(models.Model):
    name = models.CharField(max_length=200)
    email = models.EmailField()
```

```python
    def __str__(self):
        return self.name

class Entry(models.Model):
    blog = models.ForeignKey(Blog, on_delete=models.CASCADE)
    headline = models.CharField(max_length=255)
    body_text = models.TextField()
    pub_date = models.DateField()
    mod_date = models.DateField()
    authors = models.ManyToManyField(Author)
    number_of_comments = models.IntegerField()
    number_of_pingbacks = models.IntegerField()
    rating = models.IntegerField()

    def __str__(self):
        return self.headline
```

8.10.1 创建对象

在 shell 窗口执行以下代码将会创建博客对象：

```
>>> python .\manage.py shell
>>> from blog.models import Blog
>>> b = Blog(name='Beatles Blog', tagline='All the latest Beatles news.')
>>> b.save()
```

8.10.2 修改对象

Django 中修改对象非常简单，只要获取对象然后修改对象属性值即可：

```
>>> blog = Blog.objects.get(name="Beatles Blog")
>>> blog.name = "Cheddar Talk"
>>> blog.save()
```

8.10.3 更新 ForeignKey

更新模型的 ForeignKey 字段与更新其他普通字段一样，只需要直接给 ForeignKey 字段赋值即可。进行后续代码之前先添加一个 entry 对象以及一个新的博客：

```
>>> from blog.models import Blog, Entry
>>> import time
>>> entry = Entry.objects.create(blog=blog,
    pub_date=time.strftime("%Y-%m-%d",time.localtime()),
    mod_date=time.strftime("%Y-%m-%d",time.localtime()),
    number_of_comments=0,
    number_of_pingbacks=0,
    rating=0)
>>> aaron_blog = Blog.objects.create(name="Aarons Blog")
```

例如：

```
>>> from blog.models import Blog, Entry
>>> entry = Entry.objects.get(pk=1)
```

```
>>> new_blog = Blog.objects.get(name="Aarons Blog")
>>> entry.blog = new_blog
>>> entry.save()
```

代码分析：

Entry.objects.get(pk=1)：在数据库中取得主键为 1 的记录（如果不存在主键为 1 的记录将出现异常：Entry matching query does not exist）。

Blog.objects.get(name=" Aarons Blog")：在数据库中查找 name 为"Aarons Blog"的博客。

entry.blog = new_blog：更新 entry 的 blog 属性。

entry.save()：保存新 Entry 对象。

8.10.4　更新 ManyToManyField

由于 ManyToManyField 字段不像其他字段一样可以直接在数据表中保存属性值，ManyToManyField 字段需要通过一张关系表来保存所有相关联的数据，因此更新 ManyToManyField 字段的方式也与其他字段不太一样。

添加一个外键：

```
>>> from blog.models import Author
>>> joe = Author.objects.create(name="Joe")
>>> entry.authors.add(joe)
```

添加多个外键：

```
>>> john = Author.objects.create(name="John")
>>> paul = Author.objects.create(name="Paul")
>>> george = Author.objects.create(name="George")
>>> ringo = Author.objects.create(name="Ringo")
>>> entry.authors.add(john, paul, george, ringo)
```

代码分析：

以上代码前四行分别创建 4 个作者实例对象：john、paul、george、ringo。

entry.authors.add(john, paul, george, ringo)：将新建的 4 个作者实例对象添加到关联表中与 Entry 进行关联。

作者表如下图所示。

	id	name	email
1	1	Joe	
2	2	John	
3	3	Paul	
4	4	George	
5	5	Ringo	

作者 - 博客关联表如下图所示。

	id	entry_id	author_id
1	1	1	1
2	2	1	2
3	3	1	3
4	4	1	4
5	5	1	5

8.10.5 数据查询

可以直接使用 QuerySet 对象的 all() 方法查询数据表中的全部数据，如：

```
>>> all_entries = Entry.objects.all()
```

也可以使用恰当的过滤器查询特定数据，除了 all() 之外，Manager 对象还提供以下过滤器。

- filter(**kwargs)：返回一个新的 QuerySet 对象，新对象只包含符合过滤条件的数据。
- exclude(**kwargs)：返回一个新的 QuerySet 对象，新对象不包含符合过滤条件的数据。

过滤器参数 kwargs 与 SQL 脚本中的 WHERE 条件语句一样用于过滤数据，它们在 QuerySet 对象方法中以关键字参数的形式存在，书写格式为 field__lookuptype=value（注意 field 与 lookuptype 之间是两个下画线）。例如，查询发布日期早于或等于 2016 年 1 月 1 日的所有文章：

```
>>> Entry.objects.filter(pub_date__lte='2006-01-01')
```

等效的 SQL 语句如下：

```
SELECT * FROM blog_entry WHERE pub_date <= '2006-01-01';
```

通常情况，过滤语句中的 field 是模型的字段名，唯一的例外情况是 ForeignKey 字段，如果需要通过外键过滤数据，需要使用 ForeignKey 字段名 + "_id" 的形式。例如，查找所有博客 id 为 4 的 Entry 数据：

```
>>> Entry.objects.filter(blog_id=4)
```

某些情况下只需要查找唯一的一条数据，那么使用 filter 显得不是很方便。此时可以使用 get() 方法，该方法可以直接返回对象实例：

```
>>> blog = Blog.objects.get(id=1)
>>> blog
<Blog: Cheddar Talk>
>>> blog.name
'Cheddar Talk'
```

除了 all()、get()、filter() 和 exclude() 外，QuerySet 还提供了其他很多方法，下面介绍几个常用的方法，这些方法执行结束仍然返回 QuerySet 对象。

order_by(*fields)

默认情况下，QuerySet 根据模型的 Meta 类的 ordering 属性进行排序，通过调用 order_by() 方法可以重写默认的排序规则，例如：

```
Entry.objects.filter(pub_date__year=2005).order_by('-pub_date', 'headline')
```

以上查询将对 pub_date 进行降序排列，对 headline 进行升序排列。

```
Entry.objects.order_by('?')
```

以上查询结果是随机排序的，这可能会占用更多资源，消耗更长时间。

下面两个查询是等价的，都是根据外部模型的默认排序规则进行排序，如果外部模型没有指定默认排序规则则使用外部模型的主键进行排序：

```
Entry.objects.order_by('blog')
Entry.objects.order_by('blog__id')
```

如果 Blog 模型的默认排序规则是 ordering = ['name']，那么等价的查询是：

```
Entry.objects.order_by('blog__name')
```

还可以通过调用 asc() 或者 desc() 对查询表达式进行排序：

```
Entry.objects.order_by(Coalesce('summary', 'headline').desc())
```

注意

如果排序类似于多对多字段这样的字段，返回的结果可能不准确，会比真实数据多，应当避免对这类字段进行排序。

reverse()

调用 reverse() 方法可以翻转 QuerySet 的返回结果，调用两次 reverse() 方法可以使 QuerySet 的返回结果恢复正常。

一个 reverse() 方法的应用场景是提取数据的倒数某几条：

```
my_queryset.reverse()[:5]
```

需要注意的是，对 reverse() 进行切片的工作原理与 Python 切片是不一样的，reverse()[:5] 首先返回倒数第一条数据，然后返回倒数第二条数据，依次类推直到返回 5 条数据。而 Python 的切片方法 seq[-5:] 首先返回倒数第五条数据，然后返回倒数第四条数据，依次类推。

最后需要注意的是，只有当 QuerySet 具备排序条件时才可以执行 reverse() 方法。

distinct(*fields)

对 QuerySet 进行 SELECT DISTINCT 查询，该方法可以排除重复数据。

values(*fields, **expressions)

与其他 QuerySet 方法返回模型实例不同，values() 返回一个字典。字典的 Key 是数据库字段名。

例如，查看全部博客文章：

```
>>> dict = Blog.objects.values()
>>> dict
<QuerySet [{'id': 1, 'name': 'Cheddar Talk', 'tagline': 'All the latest Beatles news.'}, {'id': 2, 'name': 'Aarons Blog', 'tagline': ''}]>
>>> dict[0]['name']
'Cheddar Talk'
```

通过为 values() 方法传递关键字参数的方式还可以修改字典的 Key：

```
>>> from django.db.models.functions import Lower
>>> dict = Blog.objects.values(blog_name=Lower('name'), blog_tagline=Lower('tagline'))
>>> dict
<QuerySet [{'blog_name': 'cheddar talk', 'blog_tagline': 'all the latest beatles news.'}, {'blog_name': 'aarons blog', 'blog_tagline': ''}]>
```

注意

关键字参数的值是一个可以传递给 QuerySet.annotate() 的表达式。QuerySet.annotate() 用于修改字段的注释。

values_list(*fields, flat=False, named=False)

与 values() 方法不同的是 values_list() 返回值是一组元组：

```
>>> dict = Blog.objects.values_list('id', 'name', 'tagline')
>>> dict
<QuerySet [(1, 'Cheddar Talk', 'All the latest Beatles news.'), (2, 'Aarons Blog', '')]>
```

如果只需要在每一行数据中提取一个字段，那么可以使用 flat 参数，这样 values_list() 的返回值就是一个最简单的列表：

```
>>> dict = Blog.objects.values_list('name', flat=True)
>>> dict
<QuerySet ['Cheddar Talk', 'Aarons Blog']>
```

如果没有为 values_list() 传递任何参数，那么 QuerySet 将会以默认排序方式返回全部数据。

需要注意的是，如果使用 values() 或者 values_list() 提取一个多对多字段值的时候，将会返回多组数据：

```
>>> entry = Entry.objects.values_list('id', 'authors__name')
>>> entry
<QuerySet [(1, 'Joe'), (1, 'John'), (1, 'Paul'), (1, 'George'), (1, 'Ringo')]>
```

select_related(*fields)

根据外键关系生成 QuerySet。当执行查询时搜索相关外键对象，这对于提升性能非常有帮助，它通过生成一个复杂的查询使得以后需要查询外键时不必重复数据库操作。

除了能够返回 QuerySet 对象的方法外，QuerySet 还提供了一些不返回 QuerySet 的方法，如前面提到的 get() 方法。下面介绍一下其他不返回 QuerySet 对象的常见方法。

create(kwargs)**

创建并保存模型对象，例如：

```
john = Author.objects.create(name="John")
```

get_or_create(defaults=None, **kwargs)

通过指定的关键字查询一条数据，在必要情况下创建数据。如果所有模型字段都有默认值，那么可以不提供参数 kwargs。

该方法返回一个两位的元组（object, created），其中，object 是从数据库中提取的或者新建的对象；created 是一个布尔值，用于标识第一个返回值的对象是否是新创建的。

使用该方法可以有效地避免添加重复数据。

下面使用 get_or_create() 方法重复添加两个次名为 Aaron 的作者：

```
>>> author = Author.objects.get_or_create(
...     name="Aaron",
...     email="aaron@test.com"
... )
>>>
>>> author
(<Author: Aaron>, True)
>>> author = Author.objects.get_or_create(
...     name="Aaron",
...     email="aaron@test.com"
... )
>>>
>>> author
(<Author: Aaron>, False)
```

update_or_create(defaults=None, **kwargs)

与 get_or_create() 方法非常相似，不过 update_or_create() 方法用于更新对象，当与 kwargs 匹配的对象不存在时创建一个新的。

参数 defaults 是一个用于更新对象的字典，字典的 Key 和 value 分别是模型的字段名与字段值。defaults 也可以是一个可调用对象。

该方法返回一个两位的元组（object, created），其中，object 是更新后的对象或者新建的对象；created 是一个布尔值，用于标识第一个返回值的对象是否是新创建的。

下面使用 update_or_create() 方法更新作者 Aaron 的电子邮箱：

```
>>> author = Author.objects.update_or_create(
... name="Aaron",
... defaults={"email":"aaron@django.com"},
... )
>>>
>>> author
(<Author: Aaron>, False)
>>> author[0].email
'aaron@django.com'
```

bulk_create(objs, batch_size=None, ignore_conflicts=False)

向数据库中批量添加数据。

下面代码可以创建两个 Blog 对象：

```
>>> Blog.objects.bulk_create([Blog(name="Awsome Django"), Blog(name="MTV Framework")])
[<Blog: Awsome Django>, <Blog: MTV Framework>]
```

进入数据库，结果如下图所示。

	id	name	tagline
1	1	Cheddar Talk	All the latest Beatles news.
2	2	Aarons Blog	
3	3	Awsome Django	
4	4	MTV Framework	

使用 bulk_create 时需要注意以下几点：
- bulk_create() 不会调用模型的 save() 方法，也不会发送 pre_save 和 post_save 信号。
- 在多表继承场景中不能应用于子模型。
- 如果模型的主键是一个 AutoField 字段，那么 bulk_create 不会像 save() 方法那样检索和设置主键，除非数据库本身支持。
- 不能应用于多对多关系。
- bulk_create() 首先会将参数 objs 转换成一个列表，如果 objs 是一个生成器就会被完全执行。

在支持它的数据库上（除 Oracle 外），若 ignore_conflicts 参数设置为 True，将告诉数据库忽略插入失败的情况。启用此参数将禁用在每个模型实例上设置主键。

bulk_update(objs, fields, batch_size=None)

批量更新数据库中的数据。

下面代码同时更新两个博客的 tagline 字段：

```
>>> blog1 = Blog.objects.get(name="Awsome Django")
>>> blog2 = Blog.objects.get(name="MTV Framework")
>>> blog1.tagline = "Django web framework is awsome"
>>> blog2.tagline = "Django uses MTV framework"
>>> Blog.objects.bulk_update([blog1, blog2], ['tagline'])
```

进入数据库，结果如下图所示。

id	name	tagline
1	Cheddar Talk	All the latest Beatles news.
2	Aarons Blog	
3	Awsome Django	Django web framework is awsome
4	MTV Framework	Django uses MTV framework

使用 bulk_update 时需要注意以下几点：
- 不能更新模型主键。
- 使用 bulk_update 不会调用模型的 save() 方法，也不会发送 pre_save 和 post_save 信号。
- 为了避免一次修改过多数据造成自动生成的 SQL 语句过长，应该使用 batch_size 进行单次更新的数量限制。
- 在多表继承关系中，如果更新祖先表将会产生额外的 SQL 查询。
- 如果 objs 参数中存在重复值，那么只针对第一个进行更新。

batch_size 用于控制一次更新的数据量，默认全部更新。

count()

返回 QuerySet 中的数据量。

常用方式如下：

```
>>> Blog.objects.filter(tagline__contains='Django').count()
2
```

由于 count() 执行的 SQL 语句是 "SELECT COUNT(*)"，因此比使用 len 计算对象数量性能要好。

latest(*fields)

根据指定字段返回数据库中的最新数据。

常用方式如下：

```
>>> latest_entry = Entry.objects.latest('pub_date')
>>> latest_entry.pub_date
datetime.date(2020, 3, 5)
```

earliest(*fields)

与 latest 作用相反，取最早的值。

first()

取得 QuerySet 中第一个值，如果没有匹配项则返回 None。

last()

与 first 作用相反，取最后一个值。

exists()

以较高的性能判断 QuerySet 中是否存在匹配项。

常用方式如下：

```
entry = Entry.objects.get(pk=123)
```

```
if some_queryset.filter(pk=entry.pk).exists():
    print("Entry contained in queryset")
```

update(kwargs)**

执行 SQL update 语句更新指定字段并返回匹配的数量（该数据仅与匹配的数据量有关）。

示例代码：

```
>>> Entry.objects.filter(pub_date__year=2010).update(comments_on=False)
```

delete()

执行 SQL delete 语句删除全部匹配的 QuerySet 数据。返回被删除的对象总数以及每一类对象具体被删除的数量，例如：

```
>>> e.delete()
(1, {'weblog.Entry': 1})

>>> Entry.objects.filter(pub_date__year=2005).delete()
(5, {'webapp.Entry': 5})
```

> **注意**
>
> ❑ 当被删除的对象是其他模型数据的外键时，其他模型中相应的数据也会被删除。
> ❑ delete() 是唯一一个没有在 Manager 上暴露的数据库方法，这样可以在一定程度上防止用户不小心使用 Entry.objects.delete() 删除全部数据，如果用户确实需要删除全部数据应该使用如下方式：
>
> ```
> Entry.objects.all().delete()
> ```

8.10.6 链式过滤器

由于过滤器的返回结果是 QuerySet，因此可以对 QuerySet 对象进行链式查询，例如：

```
>>> Entry.objects.filter(
...     headline__startswith='What'
... ).exclude(
...     pub_date__gte=datetime.date.today()
... ).filter(
...     pub_date__gte=datetime.date(2005, 1, 30)
... )
```

以上查询语句首先从全部 Entry 数据中查找 headline 以"What"开头的子集，然后从子集中排除发布日期大于或等于今天的数据，最后在第二步的结果中查找发布日期大于或等于 2005 年 1 月 30 日的数据。

需要注意的是，每次通过过滤器生成的 QuerySet 都是一个全新的对象，因此可以被重复使用。

另外，QuerySet 是惰性的，也就是说，当使用过滤器创建一个 QuerySet 时，Django 仅保存查询条件，只有当真正计算时才执行查询。不同时期执行 QuerySet 可能产生不同的结果。

8.10.7 查询条件

前面讲到 QuerySet 的查询参数的书写格式为 field__lookuptype=value，其中，lookuptype 是查询条件，类似于 SQL 脚本中 WHERE 语句的比较运算符。下面是 Django 自带的查询条件。

exact

完全匹配运算符。由于完全匹配的使用频率最高，因此 Django 将 exact 定义为默认查询条件，如果在过滤语句中没有指定查询条件，那么 Django 将按照完全匹配查找数据，例如：

```
Entry.objects.get(id=14)
```

等价于：

```
Entry.objects.get(id__exact=14)
```

如果在 exact 运算符右侧指定 None，那么在翻译成 SQL 语句时就会按照 SQL 中的 NULL 进行比较。

例如下面的 Django 语句：

```
Entry.objects.get(id__exact=14)
Entry.objects.get(id__exact=None)
```

等价于如下的 SQL 语句：

```
SELECT ... WHERE id = 14;
SELECT ... WHERE id IS NULL;
```

iexact

等同于 exact 运算符，但是不区分字母大小写。

例如下面的 Django 语句：

```
Blog.objects.get(name__iexact='beatles blog')
Blog.objects.get(name__iexact=None)
```

等价于如下的 SQL 语句：

```
SELECT ... WHERE name ILIKE 'beatles blog';
SELECT ... WHERE name IS NULL;
```

上面第一个查询语句将会查找到所有包含"beatles blog"的文章，如包含"Beatles Blog""beatles blog"和"BeAtLes BLoG"的文章。

contains

包含查询，区分字母大小写。

例如查找标题包含 Django 的博客：

```
Entry.objects.get(headline__contains='Django')
```

对应的 SQL 语句如下：

```
SELECT ... WHERE headline LIKE '%Django%';
```

> **注意**
>
> 此时只会查找到标题包含 Django 的文章，而不会查找包含"django"或者"DJANGO"的文章。

icontains

等同于 contains 运算符，但是不区分字母大小写。

同样使用上面的查询语句，但是换作 icontains 过滤条件将会检索到所有包含"Django"的文章，而不关心字母的大小写形式。

in

字段值存在于一个可迭代列表中，如用列表、元组或者 QuerySet 等。等价于 SQL 的 IN 操作。
例如：

```
Blog.objects.filter(id__in=[1, 3, 5])
Entry.objects.filter(headline__in='abc')
```

等价于：

```
SELECT ... WHERE id IN (1, 3, 5);
SELECT ... WHERE headline IN ('a', 'b', 'c');
```

除此之外，in 还可以进行更复杂的对象比较，例如：

```
inner_qs = Blog.objects.filter(name__contains='Cheddar')
entries = Entry.objects.filter(blog__in=inner_qs)
```

对应的 SQL 语句如下：

```
SELECT ... WHERE blog.id IN (SELECT id FROM ... WHERE NAME LIKE '%Cheddar%')
```

in 运算符还可以和 values、values_list 结合使用，但是一定要注意此时 values 和 values_list 只能接收一个字段，例如下面的写法是合法的：

```
inner_qs = Blog.objects.filter(name__contains='Ch').values('name')
entries = Entry.objects.filter(blog__name__in=inner_qs)
```

而下面的写法就是非法的：

```
inner_qs = Blog.objects.filter(name__contains='Ch').values('name', 'id')
entries = Entry.objects.filter(blog__name__in=inner_qs)
```

在此还应注意性能问题，很多数据库没有对混合的 SQL 语句进行优化，例如以下代码可能会带来很大的性能问题：

```
SELECT * FROM Entry WHERE blog_id IN (SELECT id FROM Blog WHERE NAME LIKE '%Django%')
```

Django 推荐使用 values 方法进行查询，并且将复杂的查询拆分为多个简单查询。

gt、gte、lt、lte

分别对应 SQL 中的 >、>=、<、<=。

startswith

字段以给定值开始，区分字母大小写。

例如下面的 Django 查询语句：

```
Entry.objects.filter(headline__startswith='Lennon')
```

对应的 SQL 语句如下：

```
SELECT ... WHERE headline LIKE 'Lennon%';
```

istartswith

等同于 startswith 运算符，但是不区分字母大小写。

例如下面的 Django 查询语句：

```
Entry.objects.filter(headline__istartswith='Lennon')
```

对应的 SQL 语句如下：

```
SELECT ... WHERE headline ILIKE 'Lennon%';
```

endswith

字段以给定值结束，区分字母大小写。

例如下面的 Django 查询语句：

```
Entry.objects.filter(headline__endswith='Lennon')
```

对应的 SQL 语句如下：

```
SELECT ... WHERE headline LIKE '%Lennon';
```

iendswith

等同于 endswith 运算符，但是不区分字母大小写。

例如下面的 Django 查询语句：

```
Entry.objects.filter(headline__iendswith='Lennon')
```

对应的 SQL 语句如下：

```
SELECT ... WHERE headline ILIKE '%Lennon';
```

range

字段值出现在一个区间中。

例如：

```
import datetime
start_date = datetime.date(2005, 1, 1)
end_date = datetime.date(2005, 3, 31)
Entry.objects.filter(pub_date__range=(start_date, end_date))
```

对应的 SQL 语句如下：

```
SELECT ... WHERE pub_date BETWEEN '2005-01-01' and '2005-03-31';
```

date、year、iso_year、month、day、week_day、hour、minute、second

分别比较日期、年、ISO 8601 格式年、月、日、星期、时、分、秒，一般要结合 gt、lt 等运算符。

例如：

```
Entry.objects.filter(pub_date__date=datetime.date(2005, 1, 1))
```

```
Entry.objects.filter(pub_date__date__gt=datetime.date(2005, 1, 1))
Entry.objects.filter(pub_date__year__gte=2005)
```

quarter

对于日期（date）或者日期时间（datetime）字段，比较日期所属季节，可选值为 1、2、3、4，分别代表一年中的四个季节。

下面的 Django 语句用于查找发布于第二季度（4 月 1 日至 6 月 30 日）的所有文章：

```
Entry.objects.filter(pub_date__quarter=2)
```

time

比较日期时间（datetime）字段的时间部分，例如：

```
Entry.objects.filter(pub_date__time=datetime.time(14, 30))
Entry.objects.filter(pub_date__time__between=(datetime.time(8), datetime.time(17)))
```

isnull

对应 SQL 语句的 IS NULL 和 IS NOT NULL，可接收参数值为 True、False。例如查找发布日期为空的博客：

```
Entry.objects.filter(pub_date__isnull=True)
```

对应的 SQL 语句如下：

```
SELECT ... WHERE pub_date IS NULL;
```

regex

使用正则表达式过滤模型字段，正则表达式中的字母区分大小写。正则表达式将应用于数据库中，例如：

```
Entry.objects.get(title__regex=r'^(An?|The) +')
```

对应的 SQL 语句如下：

```
SELECT ... WHERE title REGEXP BINARY '^(An?|The) +'; -- MySQL
SELECT ... WHERE REGEXP_LIKE(title, '^(An?|The) +', 'c'); -- Oracle
SELECT ... WHERE title ~ '^(An?|The) +'; -- PostgreSQL
SELECT ... WHERE title REGEXP '^(An?|The) +'; -- SQLite
```

建议在正则表达式前使用 r'' 防止字符转义。

iregex

等同于 regex 运算符，但是不区分字母大小写。例如：

```
Entry.objects.get(title__iregex=r'^(an?|the) +')
```

对应的的 SQL 语句如下：

```
SELECT ... WHERE title REGEXP '^(an?|the) +'; -- MySQL
SELECT ... WHERE REGEXP_LIKE(title, '^(an?|the) +', 'i'); -- Oracle
SELECT ... WHERE title ~* '^(an?|the) +'; -- PostgreSQL
SELECT ... WHERE title REGEXP '(?i)^(an?|the) +'; -- SQLite
```

8.10.8 模型深度检索

SQL 查询可以通过使用 JOIN 语法跨越多张数据库表进行检索，而 Django 可以通过模型之间的关系进行深度查询。如本章示例代码中，Blog 是 Entry 的外键，如果想查询所有 Blog 名为"Beatles

Blog"的 Entry，可以使用以下代码：

```
>>> Entry.objects.filter(blog__name='Beatles Blog')
```

代码中使用两个下画线查找 blog 的 name 字段，在这里 name 是模型 Blog 的字段，而 blog 是 Entry 的字段，此时 name 与 Entry 之间通过"__"建立联系。

QuerySet 允许进行任意深度检索，同时还支持反向检索，作为反向查询的参数时，模型的名字必须使用小写字母，如 entry。以查找所有符合条件的博客为例，这些博客在 Entry 中至少有一条记录并且对应的 Entry 的 headline 字段值是 Lennon，Django 代码如下：

```
>>> Blog.objects.filter(entry__headline__contains='Lennon')
```

如果进行多层查询，中间模型没有符合条件的数据，Django 会按照 NULL 对其处理，并且不会出现任何异常。

例如，Entry 表中存在这样一条数据：它的 blog 字段指向博客表中的一条真实数据，它的 authors 字段为空。此时使用下面语句查询 Blog 时，Django 会认为 authors 的 name 字段为空，不会因为 authors 是 NULL 而抛出异常（NULL__name 是错误的）：

```
Blog.objects.filter(entry__authors__name='Lennon')
```

下面查询语句将会返回所有作者名字为空的博客：

```
Blog.objects.filter(entry__authors__name__isnull=True)
```

如果不想查找作者为空的博客，可以使用下面的查询代码：

```
Blog.objects.filter(entry__authors__isnull=False, entry__authors__name__isnull=False)
```

8.10.9 多条件查询

介绍多条件查询前，先来创建几条测试数据。

创建 4 篇博客文章，效果如下图所示。

id	name	tagline
1	Cheddar Talk	All the latest Beatles news.
2	Aarons Blog	
3	Awsome Django	Django web framework is awsome
4	MTV Framework	Django uses MTV framework

现有作者如下图所示。

id	name	email
1	Joe	
2	John	
3	Paul	
4	George	
5	Ringo	
7	Aaron	aaron@django.com

创建 Entry 记录，如下图所示。

id	headline	body text	pub date	mod date	number_	number_	rating	blog id
1	Aaron		2018-08-05	2020-03-05	0	0	0	1
2	Aaron		2019-12-21	2020-03-05	0	0	0	2
3	Joe		2020-01-09	2020-03-05	0	0	0	3
4	Joe		2020-03-05	2020-03-05	0	0	0	4

查询所有符合条件的博客，这些博客在 Entry 表中包含数据：（1）headline 是 "Aaron"；（2）pub_date 是 2019 年的 Entry。

执行查询语句，如下图所示。

```
>>> Blog.objects.filter(entry__headline__contains='Aaron', entry__pub_date__year=2019)
<QuerySet [<Blog: Aarons Blog>]>
>>>
>>> Blog.objects.filter(entry__headline__contains='Aaron').filter(entry__pub_date__year=2019)
<QuerySet [<Blog: Aarons Blog>]>
```

使用 explain() 方法查看以上两种查询条件的区别，如下图所示。

```
>>> Blog.objects.filter(entry__headline__contains='Aaron', entry__pub_date__year=2019).explain()
'3 0 0 SCAN TABLE blog_entry\n11 0 0 SEARCH TABLE blog_blog USING INTEGER PRIMARY KEY (rowid=?)'
>>>
>>> Blog.objects.filter(entry__headline__contains='Aaron').filter(entry__pub_date__year=2019).explain()
'5 0 0 SCAN TABLE blog_entry\n10 0 0 SEARCH TABLE blog_blog USING INTEGER PRIMARY KEY (rowid=?)\n13 0 0 SEARCH TABLE blog_entry AS T3 USING INDEX blog_entry_blog_id_8cd38d8b (blog_id=?)'
```

可以看到，第一个查询只执行了一次 SEARCH 查询，第二个语句执行了两次 SEARCH，这是因为第二个语句实际上创建了两次 QuerySet。

8.10.10 主键查询

为了代码书写方便，Django 提供了一个主键查询的快捷方式，前面已经用过，这个快捷方式就是 pk，在查询条件中代表当前模型的主键。以下代码是等价的：

```
Entry.objects.get(pk=1)
Entry.objects.get(id=1)
```

使用 pk 时并不限制查询条件，如可以使用 __gt 查找博客：

```
Blog.objects.filter(pk__gt=10)
```

pk 还可以用在深度检索中：

```
Entry.objects.filter(blog__pk=3)
```

8.10.11 查询条件中的 % 和 _

SQL 查询语句中的 "%" 和 "_" 有着特殊意义："%" 匹配多个字符；"_" 匹配单个字符。为了方便使用，Django 的查询语句不对这些字符进行特殊处理，例如查找文字是否包含 "%" 可以直接编写以下代码：

```
Entry.objects.filter(headline__contains='%')
```

Django 后台会将以上代码转换为如下 SQL 查询语句：

```
SELECT ... WHERE headline LIKE '%\%%';
```

使用 connection 对象可以查看当前会话中所有执行过的 SQL：

```
>>> from django.db import connection
>>> entry = Entry.objects.filter(headline__contains='%')
>>> connection.queries[-1]
{'sql': 'SELECT "blog_entry"."id", "blog_entry"."blog_id", "blog_
entry"."headline", "blog_entry"."body_text", "blog_entry"."pub_date", "blog_
entry"."mod_date", "blog_entry"."number_of_comments", "blog_entry"."number_of_
pingbacks", "blog_entry"."rating" FROM "blog_entry" WHERE "blog_entry"."headline"
LIKE \'%\\%%\' ESCAPE \'\\\' LIMIT 21', 'time': '0.000'}
```

8.10.12 F() 函数

前面示例代码中所有的查询条件都是将一个模型字段与常量进行比较，那么如何将同一个模型中的两个字段进行比较呢？为此 Django 专门提供了一个 F 表达式，F 表达式可以表示一个模型字段的值，也可以表示一个注释字段（annotated column）的值。

例如，查询所有 comments 数量大于 pingbacks 数量的博客。仍然以上面的数据库为例，更新数据库中的所有 Entry，如下图所示。

id	headline	body_tex	pub_date	mod_date	number_of_comments	number_of_pingbacks	rating	blog_id
1	1 Aaron		2018-08-05	2020-03-05	0	3	0	1
2	2 Aaron		2019-12-21	2020-03-05	1	2	0	2
3	3 Joe		2020-01-09	2020-03-05	2	1	0	3
4	4 Joe		2020-03-05	2020-03-05	3	0	0	4

创建视图：

```
from django.http import HttpResponse
from django.template.loader import render_to_string
from django.db.models import F

from .models import Blog, Author, Entry

def FSearch(request):
    entries = Entry.objects.filter(number_of_comments__gt=F('number_of_pingbacks'))

    context = {
        "entries": entries
    }

    rendered =  render_to_string("entry.html", context)
    return HttpResponse(rendered)
```

创建模板 blog/templates/entry.html：

```
{% extends "base.html" %}
{% load static %}


{% block content %}
```

```
    {% for entry in entries %}
        <h2 class="blog_head"> {{entry.blog.id}} . {{entry.blog.name}} | {{ entry.
headline }}</h2>
        <p class="blog_body">
            {{ entry.blog.tagline }}
        </p>
    {% endfor %}

{% endblock %}
```

创建 URL：

```
path(r'fsearch/', views.FSearch, name='FSearch'),
```

结果如下图所示。

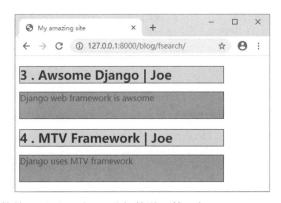

除了直接使用 F() 函数外，还可以对 F() 进行数学运算，如：

```
Entry.objects.filter(number_of_comments__gt=F('number_of_pingbacks') * 2)
Entry.objects.filter(rating__lt=F('number_of_comments') + F('number_of_pingbacks'))
```

另外在 F() 函数中也可以使用两个下画线进行深度查找：

```
Entry.objects.filter(authors__name=F('blog__name'))
```

F() 函数会持续存在，如下图所示的代码。

```
>>> from blog.models import *
>>> from django.db.models import F
>>> entry = Entry.objects.get(id=1)
>>> entry.rating
0
>>> entry.rating = F('rating') + 1
>>> entry.save()
>>> entry.body_text = "Aaron"
>>> entry.save()
>>> entry = Entry.objects.get(id=1)
>>> entry.rating
2
```

初始时，rating 值为 0，虽然只执行了一次" entry.rating = F('rating') + 1"，但是 rating 变成 2，这是因为每次保存 entry 对象时都执行了一次 F() 函数。

8.10.13　Func() 表达式

Func() 可以引用所有数据库方法，如 COALESCE、LOWER 或者其他聚合函数（如 SUM）等。

```
>>> entry = Entry.objects.annotate(blog_headline=Func(F('headline'), function=
'LOWER'))
>>> entry[0].blog_headline
'aaron'
```

代码分析：

Annotate() 方法用于生成一个注释字段，类似于 SQL 中的 as。

F('headline') 提取字段 headline。

Func() 将数据库方法 'LOWER' 应用于字段 headline 并将新值赋予字段 blog_headline。

8.10.14　QuerySet 和缓存

第一次创建 QuerySet 对象的时候，Django 不会为 QuerySet 生成任何缓存。而当第一次执行 QuerySet 的时候，Django 会缓存查询结果，后续的查询语句都可以使用当前的缓存内容。

在编写查询语句时，合理使用 QuerySet 缓存可以极大地提高代码执行效率、减少内存使用量，例如下面两条语句将会访问两次数据库并生成两个不同的 QuerySet 对象：

```
>>> print([e.headline for e in Entry.objects.all()])
>>> print([e.pub_date for e in Entry.objects.all()])
```

通常情况下，不需要这么频繁地读取数据库，只要执行一次查询操作即可，后续操作直接使用之前的查询结果就可以了，因此对以上代码进行如下修改：

```
>>> queryset = Entry.objects.all()
>>> print([p.headline for p in queryset])
>>> print([p.pub_date for p in queryset])
```

上面第二行代码是对 QuerySet 的第一次执行，此时将会缓存执行结果，当第二次执行时，操作对象就是内存中的缓存数据了，不需要继续读取数据库。

使用 connection 对象查看 QuerySet 执行了几次数据库操作，如下图所示。

```
>>> len(connection.queries)
3
>>> queryset = Entry.objects.all()
>>> print([p.headline for p in queryset])
['Aaron', 'Aaron', 'Joe', 'Joe']
>>> print([p.pub_date for p in queryset])
[datetime.date(2018, 8, 5), datetime.date(2019, 12, 21), datetime.date(2020, 1, 9), datet
ime.date(2020, 3, 5)]
>>>
>>> len(connection.queries)
4
```

在执行 Entry.objects.all() 之前，系统一共执行 3 次数据库查询，执行两次遍历操作之后，connection 对象数量只增加了 1 次。

需要注意的是，执行 QuerySet 并不是总能够生成缓存。当 QuerySet 的执行操作只影响 QuerySet 的一部分数据时，系统将不会生成缓存，这种情况通常会出现在数组的切片操作以及使用下标查询数组时，请看下面的代码：

```
>>> queryset = Entry.objects.all()
>>> print(queryset[0])
>>> print(queryset[0])
```

虽然使用 queryset[0] 执行了两次 QuerySet 查询，但是这两次操作都没有进行缓存，也就是说两次操作都直接进行了数据库查找。换句话说,如果 QuerySet 操作影响到整个查询结果,QuerySet 将会被缓存,请看以下代码:

```
>>> queryset = Entry.objects.all()
>>> [entry for entry in queryset]
>>> print(queryset[5])
>>> print(queryset[5])
```

第二行代码 [entry for entry in queryset] 遍历了整个 QuerySet,此时 Django 为查询结果创建缓存,因此后面两次 queryset[5] 操作都是从缓存中提取数据而没有直接查找数据库。

不仅遍历 QuerySet 会创建缓存,而且其他操作也会创建缓存,例如:

```
>>> [entry for entry in queryset]
>>> bool(queryset)
>>> len(queryset)
>>> list(queryset)
```

8.10.15 复杂查询与 Q 对象

前面讲的所有查询条件都是"逻辑与"运算,例如下面代码就是两个条件的"与"运算:

```
Blog.objects.filter(entry__headline__contains='Lennon', entry__pub_date__year=2017)
```

对复杂查询来说,只有"与"运算就不够了,此时可以使用 Q 对象进行复杂逻辑运算。Q 对象封装了一系列关键字参数,这些关键字参数就是前面用到的查询条件。

例如下面的 Q 对象封装了一个 LIKE 查询:

```
from django.db.models import Q
Q(question__startswith='What')
```

Q 对象之间可以使用"&"或者"|"运算符组合起来,多个 Q 对象通过运算符组合起来之后形成一个新的 Q 对象,例如下面语句将会返回一个新的 Q 对象,用于 question__startswith 之间的"逻辑或"运算。

```
Q(question__startswith='Who') | Q(question__startswith='What')
```

以上 Q 对象等价于下面的 SQL 语句:

```
WHERE question LIKE 'Who%' OR question LIKE 'What%'
```

对于复杂查询,可以使用括号对 Q 对象进行分组,另外 Q 对象可以使用波折线(~)进行取反操作,取反操作等价于 SQL 语句中的 NOT 运算,例如:

```
Q(question__startswith='Who') | ~Q(pub_date__year=2005)
```

QuerySet 中任何能够接收关键字参数的查询语句(如 filter()、exclude()、get())都可以接收 Q 对象作为位置参数。多个 Q 对象之间使用"逻辑与"运算关联,例如:

```
Poll.objects.get(
    Q(question__startswith='Who'),
    Q(pub_date=date(2005, 5, 2)) | Q(pub_date=date(2005, 5, 6))
)
```

上面的 get() 方法接收两个 Q 对象:

```
Q(question__startswith='Who')
Q(pub_date=date(2005, 5, 2)) | Q(pub_date=date(2005, 5, 6))
```

两个 Q 对象之间使用"逻辑与"运算，近似的 SQL 语句如下：

```
SELECT *
FROM polls
WHERE question LIKE 'Who%'
AND (pub_date = '2005-05-02' OR pub_date = '2005-05-06')
```

QuerySet 的查询语句可以同时接收 Q 对象和关键字参数，所有参数之间使用"逻辑与"连接。在此需要注意的一点是，如果查询语句同时接收了 Q 对象和关键字参数，那么 Q 对象一定要放置在关键字参数之前，例如：

```
Poll.objects.get(
    Q(pub_date=date(2005, 5, 2)) | Q(pub_date=date(2005, 5, 6)),
    question__startswith='Who',
)
```

8.10.16 模型比较

与 Python 一样，Django 的模型实例支持比较运算，比较运算符用"=="表示。默认情况下，模型实例是对主键进行比较，以下两种书写形式是一样的：

```
>>> some_entry == other_entry
>>> some_entry.id == other_entry.id
```

代码示例：

```
>>> b1 = Blog.objects.get(id=1)
>>> b2 = Blog.objects.get(id=2)
>>> b1 == b2
False
>>> b3 = Blog.objects.get(id=1)
>>> b1 == b3
True
```

如果模型的主键不是 id 也没有关系，Django 会自动识别主键并进行比较。

8.10.17 复制模型实例

虽然 Django 没有直接提供方法来复制模型实例，但是可以通过将已有模型实例的主键设置为 None 的简单方式创建一个完全一样的新对象实例，例如：

```
blog = Blog(name='My blog', tagline='Blogging is easy')
blog.save() # blog.pk == 1

blog.pk = None
blog.save() # blog.pk == 2
```

如果待复制的对象存在继承关系，复制对象就会比较复杂，需要将所有主键、外键都设置为 None。

8.10.18 批量更新

使用 update() 方法可以批量更新数据，例如将所有发布日期为 2007 年的博客的 headline 修改为"过期文章"：

```
Entry.objects.filter(pub_date__year=2007).update(headline='过期文章')
```

如果批量更新外键字段，只需要给外键字段重新赋值一个实例对象即可，例如：

```
>>> b = Blog.objects.get(pk=1)
>>> Entry.objects.all().update(blog=b)
```

使用 update() 更新数据库时唯一需要注意的是，update() 方法只能更新当前模型的数据表。例如，下面代码虽然可以使用关联的 Blog 对象查找主表 Entry，但是并不能更新 Blog 表，只能更新 Entry：

```
>>> b = Blog.objects.get(pk=1)
>>> Entry.objects.select_related().filter(blog=b).update(headline='Everything is the same')
```

调用 update() 方法后将会立即执行，不需要额外调用 save() 方法。

在 update() 方法中也可以使用 F 表达式，但是不能在 F 表达式中使用深度查询，例如下面的代码就是错误的：

```
>>> Entry.objects.update(headline=F('blog__name'))
```

8.10.19 模型关系

在 Django 中可以通过模型之间的关系查找数据。下面继续使用 Blog、Author、Entry 模型讲解如何利用模型之间的关系查询数据。

一对多关系

前向查询：通过 Entry 外键查询 Blog，如下图所示。

```
>>> e = Entry.objects.get(id=2)
>>> e.blog
<Blog: Aarons Blog>
```

反向查询：通过 Blog 查询相关的 Entry，如下图所示。

```
>>> b = Blog.objects.get(id=1)
>>> b.entry_set.all()
<QuerySet [<Entry: Aaron>]>
```

> **注意**
>
> entry_set 是关联模型的 Manager 实例，在关联模型中可以使用 FOO_set 格式的 Manager 实例，其中 FOO 是关联模型名字的小写格式。FOO_set 与 objects 用法完全一样。

如果认为 FOO_set 格式的 Manager 实例难以理解，可以在 Entry 类中定义 blog 外键时重写 FOO_set 名：

```
blog = ForeignKey(Blog, on_delete=models.CASCADE, related_name='entries')
```

此时调用关联模型的 Manager 实例时可以使用 'entries'：

```
>>> b = Blog.objects.get(id=1)
>>> b.entries.all() # Returns all Entry objects related to Blog.

# b.entries is a Manager that returns QuerySets.
>>> b.entries.filter(headline__contains='Lennon')
>>> b.entries.count()
```

多对多关系

多对多关系中,模型之间互相访问的方式类似于一对多关系中的反向查询。不同点在于 Manage 实例的命名:定义了 ManyToManyField 的模型使用字段名作为 Manage 实例,而"反向"模型使用源模型名的小写形式,加上 '_set'(就像反向一对多关联一样)。

```
class Author(models.Model):
    ...

class Entry(models.Model):
    authors = models.ManyToManyField(Author)
    ...
```

上面是 Entry 和 Author 模型的定义,其中,Entry 定义了 ManyToManyField 字段。此时在 Entry 中查询 Author 时可以直接使用 Entry.authors.filter() 的方式,而如果通过 Author 模型查找 Entry 必须使用 entry_set 格式。

下面是多对多模型中的查询示例:

```
e = Entry.objects.get(id=3)
e.authors.all() # Returns all Author objects for this Entry.
e.authors.count()
e.authors.filter(name__contains='John')

a = Author.objects.get(id=5)
a.entry_set.all() # Returns all Entry objects for this Author.
```

多对多关系中也可以通过修改 ManyToManyField 字段的 related_name 属性来重写关联对象的 Manager 实例名。

一对一关系

一对一关系中模型的查询方式与一对多关系中的查询方式类似。

假设存在一个模型 EntryDetail,它与 Entry 是一对一关系:

```
class EntryDetail(models.Model):
    entry = models.OneToOneField(Entry, on_delete=models.CASCADE)
    details = models.TextField()
```

可以使用以下代码查询相应模型:

```
ed = EntryDetail.objects.get(id=2)
ed.entry

e = Entry.objects.get(id=2)
e.entrydetail
```

与一对多关系的区别是反向查询的结果类型:一对多关系中查询结果是集合,而一对一关系中查询结果是模型实例,例如 e.entrydetail 是 EntryDetail 的实例。

第 9 章

视图

视图方法,简称视图,它可以接收一个 Web request 对象并向客户端返回一个 Web response 对象。response 可以是任何对象,如 HTML 文档、重定向、404 异常、XML 文档甚至一张图片。在视图方法中可以进行任意的业务逻辑处理,例如查询数据库操作等。可以将视图方法放在任何位置,只要 Python 代码能够访问到即可,通常将视图方法放在项目或者应用程序目录下的 views.py 文件中。

9.1 视图结构

下面是一个用于显示当前日期和时间的视图:

```
import datetime

from django.http import HttpResponse

def current_datetime(request):
    now = datetime.datetime.now()
    html = "<html><body>It is now %s.</body></html>" % now
    return HttpResponse(html)
```

代码分析:

导入 HttpResponse 包,该包用于生成 Web response 对象。

current_datetime 是视图方法,每一个视图方法的第一个参数都是 HttpRequest 对象,通常命名为 request。

视图方法返回 HttpResponse 对象。

9.2 HTTP 状态处理

HTTP 请求包含多种状态,如最常见的 404 错误,Django 提供了很多可以方便处理这些状态的类。例如可以使用 HttpResponseNotFound 处理 404 错误:

```
return HttpResponseNotFound('<h1>Page not found</h1>')
```

常见的 HTTP 状态码与 HttpResponse 子类的对应关系如下表所示。

HTTP 状态码	Response
301	HttpResponsePermanentRedirect
302	HttpResponseRedirect
304	HttpResponseNotModified
400	HttpResponseBadRequest
403	HttpResponseForbidden
404	HttpResponseNotFound
405	HttpResponseNotAllowed
410	HttpResponseGone
500	HttpResponseServerError

虽然 Django 提供了如此多用于处理不同 HTTP 状态的类，但是仍然不能覆盖全部可能情况，此时可以通过向 HttpResponse 传递状态码的方式来指定响应状态，例如：

```
from django.http import HttpResponse

def my_view(request):
    # ...

    # Return a "created" (201) response code.
    return HttpResponse(status=201)
```

由于 404 错误比较常见，Django 专门提供了 Http404 类用于处理它：

```
from django.http import Http404
from django.shortcuts import render

from polls.models import Poll

def detail(request, poll_id):
    try:
        p = Poll.objects.get(pk=poll_id)
    except Poll.DoesNotExist:
        raise Http404("Poll does not exist")
    return render(request, 'polls/detail.html', {'poll': p})
```

为了展示个性化的 404 错误页，可以创建一个叫作 404.html 的模板文件并把它放在模板目录的根节点，当 settings.py 中 DEBUG 属性值为 False 时，如遇到资源丢失的情况，就会调用 404.html 模板。为了方便演示，404.html 模板只包含一句话："你搜索的页面找不到了！"

Django 允许重写异常处理程序，如重写 handler404：

```
from django.conf.urls import *

def response_error_handler(request, exception=None):
    rendered = render_to_string("404.html", context)
    return HttpResponse(rendered, status=404)

handler404 = response_error_handler
```

此时访问一个不存在的页面将显示新的 404.html 模板，效果如下图所示。

9.3 快捷方式

9.3.1 render_to_string()

在前面的视图中，把 HTML 文档写在一个变量中，使用的时候将它传递给 HttpResponse 对象：

```
html = "<html><body>It is now %s.</body></html>" % now
return HttpResponse(html)
```

这样做虽然没有问题，但是，如果 HTML 文档非常大，就会导致变量内容很长，读写困难。对此首先想到的解决方案是将 HTML 文档写在文件中，在使用的时候加载到变量中，根据这个思路修改上面的代码。

步骤 1：创建一个 templates/time.html 文件。

```
<html>
    <body>
        It is now {{ time }}.
    </body>
</html>
```

步骤 2：修改视图。

```
def current_datetime(request):
    now = datetime.datetime.now()
    context = { 'time' : now }
    rendered = render_to_string('time.html', context=context)
    return HttpResponse(rendered)
```

步骤 3：添加 URL 映射。

```
path(r'time/', views.current_datetime, name='time'),
```

此时重启服务器，访问 current_datetime 视图仍然能够正常显示当前时间。

虽然使用 render_to_string() 方法已经达到目的，但是代码仍然不够简练，如果项目中包含很多视图，就需要编写很多遍类似代码。

为了提高开发速度、减少代码冗余，Django 在 django.shortcuts 包中提供了很多用于创建 Web response 对象的快捷方式。下面对这些快捷方式进行详细介绍。

9.3.2 render()

作用：将特定模板和上下文字典组合在一起并返回一个 HttpResponse 对象。

方法签名：render(request, template_name, context=None, content_type=None, status=None, using=None)

必填参数包括 request 和 template_name。各参数介绍如下。

request

HttpRequest 对象，通常是视图方法的第一个参数，用于生成 HttpResponse 对象。

template_name

一个或多个模板文件名，如果是多个模板文件名，render() 方法将选择第一个可以使用的模板进行渲染。

可选参数包括 context、content_type、status 和 using。

context

一个字典对象，字典中的元素值可以被填充到模板中，默认为空字典。如果字典值是一个可调用对象，视图方法将会在渲染模板之前调用它。

content_type

最终文档的 MIME 类型。默认值为 'text/html'。

status

HTTP 状态码，默认为 200。

using

用于加载模板的模板引擎名。

代码示例：

```
from django.shortcuts import render

def my_view(request):
    # View code here...
    return render(request, 'myapp/index.html', {
        'foo': 'bar',
    }, content_type='application/xhtml+xml')
```

以上示例等价于下面代码：

```
from django.http import HttpResponse
from django.template import loader

def my_view(request):
    # View code here...
    t = loader.get_template('myapp/index.html')
    c = {'foo': 'bar'}
    return HttpResponse(t.render(c, request), content_type='application/xhtml+xml')
```

9.3.3 redirect()

作用：返回 HttpResponseRedirect 对象，用以进行 URL 跳转。

方法签名：redirect(to, *args, permanent=False, **kwargs)

各参数介绍如下。

to

参数 to 可以是以下任意一种类型：

❏ 模型：模型的 get_absolute_url() 方法将被调用。

❏ 视图名，可能带有参数，后台使用 reverse() 方法反向解析 URL。

❑ 绝对 URL 或相对 URL，将按原样用于重定向。

permanent

默认为临时重定向，设置 permanent=True 将进行永久重定向。

下面以视图形式演示如何进行 URL 跳转。

现有 url 配置：

```
app_name = 'blog'

urlpatterns = [
    path(r'<int:blog_id>/', views.search_blog, name='detail'),
    path(r'time/', views.current_datetime, name='time'),
]
```

修改 current_datetime 视图，使得当用户访问 current_datetime 时页面跳转到 id 为 1 的博客详情页：

```
def current_datetime(request):
    return redirect("blog:detail", blog_id=1)
```

代码分析：

由于 redirect() 方法使用 reverse() 反向解析 URL，因此第一个参数应该是视图对应的 URL 名字，当系统存在多个名字为 detail 的 URL 时，应添加应用程序名，即 URL 的命名空间。

由于视图 search_blog 需要接收一个必选参数 blog_id，因此在调用 redirect() 方法时给出 blog_id=1。

以传递模型的方式修改以上代码：

```
def current_datetime(request):
    obj = Blog.objects.get(id=1)
    return redirect(obj)
```

以传递 URL 的方式修改以上代码：

```
def current_datetime(request):
    return redirect("/1/")
```

9.3.4 get_object_or_404()

作用：从模型中提取数据，如果数据不存在则调用 Http404 异常视图。

方法签名：get_object_or_404(klass, *args, **kwargs)

必填参数介绍如下。

klass

一个模型类、Manager 或者 QuerySet 对象实例。

**kwargs

查询参数，可以用于 get() 或者 filter() 方法。

以 search_blog 视图为例，验证 get_object_or_404() 方法：

```
from django.http import HttpResponse
from django.shortcuts import get_object_or_404

def search_blog(request, blog_id):
```

```
    obj = get_object_or_404(Blog, pk=blog_id)

    context = {
        "blog": obj
    }

    rendered = render_to_string("detail.html", context)
    return HttpResponse(rendered)
```

当访问一个不存在的博客时将弹出 404 错误页，如下图所示。

9.3.5　get_list_or_404()

作用：使用 filter() 方法从模型中提取一组数据，如果数据不存在则抛出 Http404 异常。

方法签名：get_list_or_404(klass, *args, **kwargs)

必填参数介绍如下。

klass

一个模型类、Manager 或者 QuerySet 对象实例。

**kwargs

查询参数，可以用于 get() 或者 filter() 方法。

以 index 视图为例，验证 get_list_or_404() 方法：

```
from django.http import HttpResponse
from django.shortcuts import get_list_or_404

def index(request):
    blogs = get_list_or_404(Blog)

    context = {
        "blogs": blogs
    }

    rendered = render_to_string("index.html", context)
    return HttpResponse(rendered)
```

9.4　视图装饰器

视图装饰器是一系列视图方法的属性，用于提供对 HTTP 请求报文的设置。

9.4.1 HTTP 方法装饰器

HTTP 方法装饰器用于约束访问视图的请求类型，该装饰器位于 django.views.decorators.http 模块。当访问视图的请求类型不正确时，HTTP 方法装饰器将会返回 django.http.HttpResponseNotAllowed 异常错误。

代码示例：

```
from django.views.decorators.http import require_http_methods

@require_http_methods(["GET", "POST"])
def my_view(request):
    # I can assume now that only GET or POST requests make it this far
    # ...
    Pass
```

需要注意的是，HTTP 请求类型必须使用大写字母。

如果仅允许使用 GET、POST 或者其他安全类型（如 GET 和 HEAD 方法），可以使用 django.views.decorators.http 模块下面的其他装饰器。

- require_GET()：装饰器可以要求视图只接受 GET 方法。
- require_POST()：装饰器可以要求视图只接受 POST 方法。
- require_safe()：装饰器可以要求视图只接收 GET 和 HEAD 方法。

9.4.2 GZip 压缩

GZip 是目前 Internet 上非常流行的数据压缩格式，对纯文本文件来说，GZip 压缩效果非常明显，可以减少 60% 至 70% 的文件大小。当用户访问网页时，Web 服务器使用 GZip 算法对网页内容进行压缩，然后将压缩后的内容传输到客户端浏览器。由于需要传递的字节数大大减少，因此网页的访问速度也会得到改善。

Django 中的 GZip 视图装饰器位于 django.views.decorators.gzip 模块。GZip 装饰器还会设置相应的 HTTP Vary 头信息。

下面以 blog 应用的 index 视图为例演示 GZip 压缩，比较压缩前后网络请求的变化。

下图是应用 GZip 压缩前的响应情况，可以看到响应长度为 896 字节。

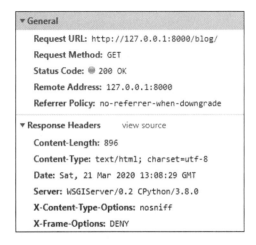

修改 index 视图，添加 GZip 装饰器：

```
from django.views.decorators.gzip import gzip_page

@gzip_page
def index(request):
    ...
```

重新访问博客首页，查看网络响应情况，如下图所示。

```
▼ General
    Request URL: http://127.0.0.1:8000/blog/
    Request Method: GET
    Status Code: ● 200 OK
    Remote Address: 127.0.0.1:8000
    Referrer Policy: no-referrer-when-downgrade
▼ Response Headers    view source
    Content-Encoding: gzip
    Content-Length: 380
    Content-Type: text/html; charset=utf-8
    Date: Sat, 21 Mar 2020 13:14:45 GMT
    Server: WSGIServer/0.2 CPython/3.8.0
    Vary: Accept-Encoding
    X-Content-Type-Options: nosniff
    X-Frame-Options: DENY
```

从上图可见，同样的请求响应长度减少到 380 字节，效果还是非常明显的，与此同时响应报文头增加了两个属性：

```
Content-Encoding: gzip
Vary: Accept-Encoding
```

9.4.3　Vary

Vary 是一个 HTTP 响应头，它决定对未来的一个请求是使用缓存还是向源服务器发起一个新的请求。一个简单的 Vary 可以像下面这样：

```
Vary: Accept-Encoding
Vary: Accept-Encoding,User-Agent
Vary: X-Some-Custom-Header,Host
Vary: *
```

通俗来讲，Vary 决定了哪些 HTTP Header 会被用来检验页面是否被缓存过。如果在 Vary 中设置了 User-Agent，那么即使用户使用移动浏览器访问过某一网络资源并生成了缓存，但是当用户改用桌面浏览器访问同一资源时，仍然会重新请求服务器，这是因为两次请求的 User-Agent 不同，服务器认为请求的资源也不同。

Django 中关于 Vary 可用的装饰器包括 vary_on_cookie 和 vary_on_headers。

下面修改 search_blog 视图，使得服务器可以缓存博客文章：

```
from django.views.decorators.vary import vary_on_headers

@vary_on_headers('Accept-Encoding')
def search_blog(request, blog_id):
    ...
```

除了直接为视图添加装饰器外,还可以使用 patch_vary_headers() 方法设置 HttpResponse 对象,例如:

```
from django.shortcuts import render
from django.utils.cache import patch_vary_headers

def my_view(request):
    ...
    response = render(request, 'template_name', context)
    patch_vary_headers(response, ['Cookie'])
    return response
```

9.4.4 缓存

缓存装饰器位于 django.views.decorators.cache 模块,用于设置服务器端和客户端缓存。

1. cache_control(**kwargs)

设置浏览器响应的 Cache-Control 头,可选参数包括下表所示的几种。

cache-directive	说明
public	表示任意响应内容都可能被缓存到任何位置,这些响应内容包括本来不应该缓存的内容或者应该被缓存在私有缓存区的内容
private	表示全部或部分响应内容都会被当作单独用户所使用,并且只缓存到私有缓存中(仅客户端可以缓存,代理服务器不可缓存)
no-cache	如果没有为 no-cache 提供字段,那么后续请求将不会使用现有的缓存,后续请求必须重新进行服务器验证 如果为 no-cache 提供了一个或多个字段,那么后续请求将会使用现有的缓存,但是缓存不包含前面提供的字段
no-store	no-store 指令可以防止由于疏忽而发布或保存敏感信息,对 HTTP 请求或响应信息有效。请求和响应都禁止被缓存
max-age=xxx	指示客户端只能接受有效期小于指定时间的响应,单位为秒,这个选项只在 HTTP 1.1 中可用

代码示例:

```
from django.views.decorators.cache import cache_control

@cache_control(private=True, max_age=3600)
def Index(request):
    blogs = Blog.objects.all()

    context = {
        "blogs": blogs
    }

    rendered = render_to_string("index.html", context)
    return HttpResponse(rendered)
```

2. never_cache(view_func)

这个装饰器添加 Cache-Control: max-age=0, no-cache, no-store, must-revalidate 头到一个响应来标识禁止缓存该页面。

代码示例:

```python
from django.views.decorators.cache import never_cache

@never_cache
def search_blog(request, blog_id):
    ...
```

使用浏览器调试工具查看 Response Headers,如下图所示。

```
▼ General
    Request URL: http://127.0.0.1:8000/blog/1/
    Request Method: GET
    Status Code: ● 200 OK
    Remote Address: 127.0.0.1:8000
    Referrer Policy: no-referrer-when-downgrade
▼ Response Headers    view source
    Cache-Control: max-age=0, no-cache, no-store, must-revalidate, private
    Content-Length: 485
    Content-Type: text/html; charset=utf-8
    Date: Sat, 21 Mar 2020 14:11:45 GMT
    Expires: Sat, 21 Mar 2020 14:11:45 GMT
    Server: WSGIServer/0.2 CPython/3.8.0
    Vary: Accept-Encoding
    X-Content-Type-Options: nosniff
    X-Frame-Options: DENY
```

关于更多 cache 指令介绍可参阅 RFC 2616,网址为 https://tools.ietf.org/html/rfc2616#section-14.9。

9.5 Django 内置视图

9.5.1 serve

为了方便开发人员调试代码,Django 预先设置了一个 serve 视图。serve 视图可以用来查看任意路径下的文件。例如,当用户上传文件后,使用 serve 视图查看文件是否保存成功。

serve 视图的定义如下:

```
static.serve(request, path, document_root, show_indexes=False)
```

可以直接在 URLconf 中调用 serve 视图,例如:

```python
from django.urls import re_path
from django.views.static import serve

# 在这里放置其他URL

if settings.DEBUG:
    urlpatterns += [
        re_path(r'^media/(?P<path>.*)$', serve, {
            'document_root': '文件夹',
        }),
    ]
```

此时所有保存在 document_root 路径下的文件都可以直接使用"/media/ 文件名"的方式进行访问。

例如，在 document_root 路径下有一个名为"Hello Django.txt"的文本文件，此时访问"http://127.0.0.1:8000/blog/media/Hello Django.txt/"的效果如下图所示。

9.5.2 错误视图

异常处理是开发人员一直需要进行的任务。由于 HTTP 异常是固定的几种，因此 Django 提前对这些异常处理进行了封装。

1. HTTP 404 视图

当视图程序抛出 HTTP 404 异常时，Django 会调用一个视图去处理它，默认视图是 django.views.defaults.page_not_found()。page_not_found() 会在网页中输出简单的"Not Found"字样或者加载 404.html。

page_not_found() 视图的定义如下：

```
defaults.page_not_found(request, exception, template_name='404.html')
```

使用 page_not_found() 视图时需要注意以下几点：
- 当 Django 无法找到匹配的 URL 时会抛出 404 错误。
- HTTP 404 视图可以接收模板上下文中的变量。
- 当 DEBUG 设置为 True 时 HTTP 404 视图将被禁用。

2. HTTP 500 视图

当出现运行时异常时，Django 会调用 django.views.defaults.server_error 视图。server_error 视图会在网页中输出简单的"Server Error"字样或者加载 500.html。

> **注意**
>
> 默认的 HTTP 500 视图不会向 500.html 传递任何变量，当 DEBUG 设置为 True 时 HTTP 500 视图将被禁用。

server_error 视图的定义如下：

```
defaults.server_error(request, template_name='500.html')
```

3. HTTP 403 视图

对于 HTTP 403 异常，默认的视图是 django.views.defaults.permission_denied，该视图会在网页中输出"403 Forbidden"或者加载 403.html。

permission_denied 视图的定义如下：

```
defaults.permission_denied(request, exception, template_name='403.html')
```

4. HTTP 400 视图

当出现 SuspiciousOperation 异常并且代码中没有进行处理时，Django 会发生"bad request"异常。默认处理"bad request"请求的视图是 django.views.defaults.bad_request。

bad_request 异常与 server_error 异常非常相似，但是 bad_request 主要是由于用户操作导致的异常，另外与 bad_request 异常相关的信息也不会被传递给模板，这样可以保证不会泄露敏感信息。

bad_request 视图要求 DEBUG=False。

9.6 HttpRequest 对象

当请求网页时，Django 会自动创建一个 HttpRequest 对象，这个对象包含所有请求中的必要数据。每一个 Django 视图都会在第一个参数位置接收 HttpRequest 对象。

9.6.1 属性

除特殊说明外，所有的 HttpRequest 对象属性都是只读的，下面是全部 HttpRequest 对象属性。

1. HttpRequest.scheme
表示请求所用网络协议，通常是 HTTP 或者 HTTPS。

2. HttpRequest.body
HTTP 请求的 body 部分，是 bytestring 类型。

3. HttpRequest.path
请求资源的全路径，如"/music/bands/the_beatles/"，不包含网络协议或者域名。

4. HttpRequest.path_info
在某些 Web 服务器上，URL 中主机名后面的内容会被拆分为两部分：前缀和 path_info。

例如，部署应用程序时，如果将 WSGIScriptAlias 设置为"/minfo"，那么 HttpRequest.path 就可能是"/minfo/music/bands/the_beatles/"，而 HttpRequest.path_info 就可能是"/music/bands/the_beatles/"。

5. HttpRequest.method
HTTP 请求所使用的方法，属性值必须大写，如 GET、POST。

6. HttpRequest.encoding
表示当前的编码类型，用于解码表单提交的数据（取值为 None 时表示使用配置文件中的 DEFAULT_CHARSET 配置项）。HttpRequest.encoding 是可编辑属性，当请求所提交的数据与 DEFAULT_CHARSET 不一致时，可以通过修改属性值的方式保证能够正确取得请求数据。

7. HttpRequest.content_type
表示 MIME 类型的字符串，取值来自 CONTENT_TYPE 头。

8. HttpRequest.content_params
字典格式，表示 CONTENT_TYPE 头的值。

例如：<meta http-equiv="content-type" content="text/html;charset=utf-8">。

9. HttpRequest.GET
类似字典类型的对象，包含所有 HTTP GET 参数。

10. HttpRequest.POST
类似字典类型的对象，包含所有 HTTP POST 参数，通常是表单数据。如果需要访问原始表单数据，可以使用 HttpRequest.body 属性。

由于网络请求允许发送空白表单，因此不能使用 if request.POST 去判断当前请求类型是否是 POST 请求，最好使用 if request.method == "POST" 来判断请求类型。

注意 POST 不包含上传的文件信息。

11. HttpRequest.COOKIES
字典对象，用于保存所有 Cookie，字典的 Key 和 Value 都是字符串。

12. HttpRequest.FILES
字典对象，用于保存所有被上传的文件。字典的 Key 是 HTML 元素 <input type="file" name="" /> 的 name，字典的值是 UploadedFile 对象。

只有当请求方式是 POST，同时表单包含 enctype="multipart/form-data" 属性的时候 HttpRequest.FILES 才包含数据，否则 HttpRequest.FILES 就是一个空的类似字典的对象。

13. HttpRequest.META
字典对象，包含所有 HTTP 头。下面是一些常用的 Header。

CONTENT_LENGTH：Request body 的长度。
CONTENT_TYPE：Request body 的 MIME 类型。
HTTP_ACCEPT：HTTP response 可以接收的文档类型。
HTTP_ACCEPT_ENCODING：HTTP response 可以接收的文档编码类型。
HTTP_ACCEPT_LANGUAGE：HTTP response 可以接收的文档语言。
HTTP_HOST：客户端发送的 HTTP Host header。
HTTP_USER_AGENT：客户端的 user-agent。
QUERY_STRING：URL 中的查询字符串，通常是"?"后面的部分。

14. HttpRequest.headers
类似于字典的不区分字母大小写的对象，它提供对请求中所有 http 前缀的头（另外加上 CONTENT_TYPE 和 CONTENT_LENGTH）的访问。

15. HttpRequest.resolver_match
ResolverMatch 实例，用于表示解析后的 URL。

例如，http://127.0.0.1:8000/blog/1/ 的 resolver_match 值如下：

```
ResolverMatch(func=blog.views.search_blog, args=(), kwargs={'blog_id': 1}, url_name=detail, app_names=['blog'], namespaces=['blog'], route=blog/<int:blog_id>/)
```

9.6.2 中间件属性

某些 Django 中间件也会为 HttpRequest 对象设置属性，当 request 对象缺少这类属性时，可以检查对应的中间件是否没有添加到 MIDDLEWARE 配置中。下面是部分中间件添加的属性。

1. HttpRequest.session

SessionMiddleware 提供的用于存储当前 session 信息的属性，属性值是一个类似字典的对象，属性值可以修改。

2. HttpRequest.site

CurrentSiteMiddleware 提供的用于存储当前网站信息的属性。属性值是 get_current_site() 方法返回的 Site 或者 RequestSite 对象。

3. HttpRequest.user

AuthenticationMiddleware 提供的用于存储当前用户的属性。属性值是 AUTH_USER_MODEL 的实例对象。如果当前没有用户登录，属性值是 AnonymousUser 对象。

在代码中可以使用 is_authenticated 判断用户是否登录，例如：

```
if request.user.is_authenticated:
    ...
else:
    ...
```

9.6.3 方法

1. HttpRequest.get_host()

根据 HTTP_X_FORWARDED_HOST 和 HTTP_HOST 头的值取得原始请求的主机名。如果这两个 Header 都没有值，get_host() 方法返回 SERVER_NAME + SERVER_PORT，如 127.0.0.1:8000。

2. HttpRequest.get_port()

根据 HTTP_X_FORWARDED_PORT 头和名为 SERVER_PORT 的 META 属性返回请求的原始端口号。

3. HttpRequest.get_full_path()

返回 path 属性以及查询字符串（如果有），如 /music/bands/the_beatles/?print=true。

4. HttpRequest.get_full_path_info()

返回 path_info 属性以及查询字符串（如果有），如 /minfo/music/bands/the_beatles/? print=true。

5. HttpRequest.build_absolute_uri(location=None)

从 localtion 位置开始返回绝对 URI，如果没有给定 location，默认会使用 HttpRequest.get_full_path() 替代 location。

如果 location 已经是一个绝对 URL，那么输出结果不变，否则使用 request 对象内部变量生成一个绝对 URI。

接下来以访问 http://127.0.0.1:8000/blog/1/ 为例演示 build_absolute_uri() 在不同情况下的返回值，如下图所示。

```
request.get_full_path()
'/blog/1/'
request.build_absolute_uri()
'http://127.0.0.1:8000/blog/1/'
request.build_absolute_uri(request.build_absolute_uri())
'http://127.0.0.1:8000/blog/1/'
request.build_absolute_uri('/blog/')
'http://127.0.0.1:8000/blog/'
request.build_absolute_uri('http://127.0.0.1:8000/blog/1/')
'http://127.0.0.1:8000/blog/1/'
```

6. HttpRequest.get_signed_cookie(Key, default=RAISE_ERROR, salt='', max_age=None)

返回一个已签名的 Cookie 的值，如果签名已失效则抛出异常。

如果调用方法时指定了 default 的值，那么默认的异常信息将会被 default 值替代。

在调用方法时给出 salt 值，可以有效防止网站暴力破解攻击。例如：

```
request.get_signed_cookie('name', salt='name-salt')
```

7. HttpRequest.is_secure()

如果网站启用了 HTTPS 协议，is_secure() 方法返回 True，否则返回 False。

8. HttpRequest.is_ajax()

如果请求是通过 XMLHttpRequest 对象发送的，is_ajax() 方法返回 True，否则返回 False。该方法通过检查 HTTP_X_REQUESTED_WITH 头是否包含字符串 "XMLHttpRequest" 来确定客户端的请求方式。大多数 JavaScript 类库都支持该报文头。如果自己编写 XMLHttpRequest 请求，一定要记得正确添加 HTTP_X_REQUESTED_WITH 头，否则可能导致 is_ajax() 方法不能正确识别请求方式。

9.6.4　QueryDict 对象

前面多次提到 GET 和 POST 属性是 "类似字典的对象"，其实这就是一个 django.http.QueryDict 对象实例。QueryDict 与普通字典对象最大的区别就是，QueryDict 对象允许一个 Key 对应多个 Value。

正常情况下 QueryDict 是不可变的，使用 QueryDict.copy() 方法可以得到一个可变版本的 QueryDict 对象，如下图所示。

```
qd = request.GET

qd = request.GET.copy()

qd
<QueryDict: {}>
```

由于 QueryDict 是字典的子类，因此实现了字典的所有方法。下面是 QueryDict 额外提供的方法。

1. QueryDict.__init__(query_string=None, mutable=False, encoding=None)

QueryDict 的构造方法，例如：

```
>>> QueryDict('a=1&a=2&c=3')
<QueryDict: {'a': ['1', '2'], 'c': ['3']}>
```

2. QueryDict.fromKeys(iterable, value='', mutable=False, encoding=None)

使用一个可迭代对象创建一个 QueryDict 实例。

```
>>> QueryDict.fromKeys(['a', 'a', 'b'], value='val')
<QueryDict: {'a': ['val', 'val'], 'b': ['val']}>
```

3. QueryDict.__getitem__(Key)

返回指定 Key 的值，如果 Key 包含多个值则返回最后一个值；如果 Key 不存在，则抛出 django.utils.datastructures.MultiValueDictKeyError 异常。

4. QueryDict.__setitem__(Key, value)

为 Key 赋值，新值为 [value]，如下图所示。

```
qd.__setitem__('name', 'Aaron')
None
qd.__setitem__('name', 'John')
None
qd
<QueryDict: {'name': ['John']}>
```

5. QueryDict.__contains__(Key)

判断 Key 是否存在，如下图所示。

```
qd.__contains__("name")
True
qd.__contains__("age")
False
```

由于该方法的存在使得判断 Key 更加简单，例如：

```
if "foo" in request.GET:
    ...
```

9.7　HttpResponse 对象

HttpResponse 对象是对用户访问的响应，与 HttpRequest 对象不同的是，HttpResponse 对象需要开发人员在视图中创建。

HttpResponse 对象属于 django.http 模块，可以通过向构造函数传递网页内容的方式来构造 HttpResponse 实例，在实例化的同时可以指定浏览器对文本的处理方式，例如：

```
>>> from django.http import HttpResponse
>>> response = HttpResponse("Here's the text of the Web page.")
>>> response = HttpResponse("Text only, please.", content_type="text/plain")
>>> response = HttpResponse(b'Bytestrings are also accepted.')
>>> response = HttpResponse(memoryview(b'Memoryview as well.'))
```

如果文本内容过长，还可以像文件对象一样，将文本分批写入，例如：

```
>>> response = HttpResponse()
>>> response.write("<p>Here's the text of the Web page.</p>")
>>> response.write("<p>Here's another paragraph.</p>")
```

另外可以直接操作 HttpResponse 的 Header 信息，例如：

```
>>> response = HttpResponse()
>>> response['Age'] = 120
>>> del response['Age']
```

注意

与字典不同的是，如果删除一个不存在的 Key，del 方法并不会抛出异常。

HttpResponse 构造函数还可以接收迭代器，此时将会立即消费迭代器内容；如果迭代器包含 close 方法，当消费完则立即关闭迭代器。

下面视图方法使用迭代器生成 HttpResponse。

```
def test_iterator(request):
    content = (str(i)+" " for i in range(10))
    return HttpResponse(content)
```

输入网址，结果如下图所示。

9.7.1 属性

1. HttpResponse.content

HTTP 响应的内容。

2. HttpResponse.charset

HTTP 响应所使用的编码格式。如果在构建 HttpResponse 对象的时候没有给出 charset，则会从 content_type 中提取字符编码格式，提取失败时使用配置文件中的 DEFAULT_CHARSET。

3. HttpResponse.status_code

HTTP 响应的状态码。除非显式设置 HttpResponse.reason_phrase，否则在 HttpResponse 构造函数外对 HttpResponse.status_code 的修改也会改变 HttpResponse.reason_phrase。

4. HttpResponse.reason_phrase

W3C 定义的 Reason-Phrases，每一个 HTTP 状态码都对应一个 Reason-Phrases 字符串，如下表所示。

状态码	Reason-Phrases
100	Continue
200	OK
403	Forbidden
404	Not Found
500	Internal Server Error

关于 HTTP Status Code 与 Reason-Phrases 更多信息请参阅 https://urivalet.com/reason-phrases。

5. HttpResponse.streaming

该属性值永远为 False。

由于 HttpResponse.streaming 属性的存在，Django 中间件才可以使用不同的方式来处理流响应。

6. HttpResponse.closed

如果响应已经关闭则返回 True。

9.7.2 方法

1. HttpResponse.__init__(content=b'', content_type=None, status=200, reason=None, charset=None)

使用指定的文档内容和文档类型构造 HttpResponse 对象。

各参数介绍如下。

content：迭代器、bytestring、memoryview 或者字符串。
content_type：可选的 MIME 类型，用于填充 HTTP 内容类型头 Content-Type。
status: HTTP 状态码，为了方便理解，可以使用 http.HTTPStatus 对象。
reason：HTTP Reason-Phrases。
charset：HttpResponse 所使用的编码字符集。

2. HttpResponse.__setitem__(header, value)

设置 HTTP 头，header 和 value 都是字符串。

3. HttpResponse.__delitem__(header)

删除指定 HTTP 头，如果删除失败不会抛出异常，不区分字母大小写。

4. HttpResponse.__getitem__(header)

返回指定 HTTP 头的值，不区分字母大小写。

5. HttpResponse.has_header(header)

判断 HTTP 头是否存在，不区分字母大小写。

6. HttpResponse.setdefault(header, value)

如果指定的 HTTP 头还没有设置则进行设置。

7. HttpResponse.set_cookie(Key, value='', max_age=None, expires=None, path='/', domain=None, secure=False, httponly=False, samesite=None)

设置一个 Cookie。

各参数介绍如下。

max_age：Cookie 的最长生命周期，单位为秒。默认值为 None，此时 Cookie 的生命周期与浏览器的 session 一样。当 expires 参数没有赋值的时候，才计算 max_age。

expires：Cookie 的过期时间，格式为 "Wdy, DD-Mon-YY HH:MM:SS GMT" 的字符串或者 UTC 格式的 datetime.datetime 对象。当 expires 是 datetime 对象时，计算 max_age。

domain：Cookie 所在域。当设置 domain="example.com" 时，www.example.com 和 blog.example.com 等域都可以读取该 Cookie，否则只能被当前域读取。

当设置 secure=True 时，只有 HTTPS 请求才能将 Cookie 发送到服务器。

httponly=True：阻止客户端 JavaScript 访问 Cookie。

使用 samesite='Strict' 或者 samesite='Lax' 可以告诉浏览器当进行跨源请求（cross-origin request）时不发送 Cookie。

8. HttpResponse.set_signed_cookie(Key, value, salt='', max_age=None, expires=None, path='/', domain=None, secure=False, httponly=False, samesite=None)

与 set_cookie() 方法相似，不过 set_signed_cookie() 方法在设置 Cookie 之前会进行加密。可以使用 salt 增加 Key 的安全性，但是需要注意将 salt 传递给对应的 HttpRequest.get_signed_cookie() 方法。

9. HttpResponse.delete_cookie(Key, path='/', domain=None)

删除 Cookie。删除失败不会抛出异常。

除了以上方法外，HttpResponse 对象还可以像文件或者流一样进行读写操作，具体方法如下。

- HttpResponse.close()：请求结束，自动被 WSGI 服务器调用。
- HttpResponse.write(content)：像文件对象一样写。
- HttpResponse.flush()：像文件对象一样将所有缓冲的数据强制发送到目的地。
- HttpResponse.tell()：像文件对象一样返回文件的当前位置。
- HttpResponse.getvalue()：像流对象一样返回 HttpResponse.content 的值。
- HttpResponse.readable()：永远是 False。
- HttpResponse.seekable()：永远是 False。
- HttpResponse.writable()：永远是 True。
- HttpResponse.writelines(lines)：向响应写入一些行信息。

9.7.3　HttpResponse 子类

为了处理不同类型的 HTTP 响应，Django 还提供了一些 HttpResponse 子类，分别如下。

- class HttpResponseRedirect：将请求跳转到其他地址。对应的 HTTP 状态码是 302。
- class HttpResponsePermanentRedirect：与 HttpResponseRedirect 相似，需要进行页面跳转，不过 HTTP 状态码是 301。
- class HttpResponseNotModified：表示从用户最后一次访问到现在页面都没有发生改变，HTTP 状态码是 304。
- class HttpResponseBadRequest：HTTP 状态码是 400。
- class HttpResponseNotFound：HTTP 状态码是 404。
- class HttpResponseForbidden：HTTP 状态码是 403。
- class HttpResponseNotAllowed：HTTP 状态码是 405。构造函数的第一个参数是必需的，参数值是一组任意允许的 HTTP method，如 ['GET', 'POST']。
- class HttpResponseGone：HTTP 状态码是 410。
- class HttpResponseServerError：HTTP 状态码是 500。

9.7.4　JsonResponse

JsonResponse 也是 HttpResponse 子类，用于生成 JSON 格式的响应，它的 Content-Type 头 header 默认是 application/json。

代码示例：

```
>>> from django.http import JsonResponse
>>> response = JsonResponse({'foo': 'bar'})
>>> response.content
b'{"foo": "bar"}'
```

9.7.5 FileResponse

FileResponse 是 StreamingHttpResponse 的子类，优化了对二进制文件的处理。

当参数 as_attachment 设置为 True 时，Content-Disposition 报文头将被设置为 attachment，浏览器将为用户下载文件，否则 Content-Disposition 报文头将被设置为 inline。

代码示例：

```
def test_file(request):
    return FileResponse(open(r'D:/Code/django3/mysite/blog/测试.txt','rb'), as_attachment=True,charset="utf-8")
```

 注意

只能使用二进制方法读取文件（'rb'）。

9.8 TemplateResponse 对象

由于 HttpResponse 对象在初始化结束后文档内容已经固定，很难进行修改，因此在使用中可能会遇到一些不便，例如修改 HttpResponse 对象所使用的模板，或者在现有模板中添加新数据，这些都很难实现。为了解决这些问题，Django 提供了一个全新的对象：TemplateResponse。与 HttpResponse 不同的是，TemplateResponse 会保留模板和上下文对象，直到需要输出时才将模板和上下文对象编译成 response 并返回。

9.8.1 SimpleTemplateResponse 对象

1. SimpleTemplateResponse 属性

SimpleTemplateResponse 是 TemplateResponse 的基类，包含以下属性。

- SimpleTemplateResponse.template_name。SimpleTemplateResponse 对象所使用的模板，可接收的参数包括：模板对象（如使用 get_template() 方法返回的模板对象）、单个模板名、一组模板名。例如：['foo.html', 'path/to/bar.html']。
- SimpleTemplateResponse.context_data。渲染模板时所使用的上下文对象，必须是字典类型。例如：{'foo': 123}。
- SimpleTemplateResponse.rendered_content。使用当前的模板以及上下文对象所渲染的 HTML 文档内容。
- SimpleTemplateResponse.is_rendered。布尔值，表示当前 HTML 文档对象是否已经渲染完成。

2. SimpleTemplateResponse 方法

SimpleTemplateResponse 对象包含以下多个方法。

```
SimpleTemplateResponse.__init__(template, context=None, content_type=None, status=None, charset=None, using=None)
```

使用给定的模板、上下文对象、文档类型、HTTP 状态码、字符集初始化 SimpleTemplateResponse 对象。

各参数介绍如下。
- template：可以是模板对象（如使用 get_template() 方法返回的模板对象）、单个模板名、一组模板名。
- context：渲染模板时所使用的上下文对象，字典类型。默认为 None。
- content_type：用于指定 HTTP 文档的 Content-Type 头，包括 MIME 类型和字符集。如果初始化构造函数时给出了 content_type，则使用给定的 content_type，否则使用 'text/html'。
- status：响应的 HTTP 状态码。
- charset：响应文档所使用的字符集编码格式。如果没有指定 charset，则根据 content_type 设置 charset，如果仍然无法确认 charset，则使用配置文件中的 DEFAULT_CHARSET 参数。
- using：模板引擎名。

```
SimpleTemplateResponse.resolve_context(context)
```

上下文对象的预处理方法，接收一个字典类型的上下文对象。默认返回相同的字典。
重写该方法可以自定义上下文对象。

```
SimpleTemplateResponse.resolve_template(template)
```

为页面渲染解析模板实例。该方法可以接收模板对象（如使用 get_template() 方法返回的模板对象）、单个模板名、一组模板名。
该方法返回的模板实例用于渲染网页。
重写该方法可以自定义模板的加载过程。

```
SimpleTemplateResponse.add_post_render_callback()
```

为模板渲染程序添加回调函数。通过回调函数可以有效推迟默认进程的执行，如使用回调函数可以保证只有模板渲染结束之后才可以执行缓存程序。
当 SimpleTemplateResponse 对象渲染结束，回调函数将会被立即执行。回调函数只能接收一个 SimpleTemplateResponse 对象实例作为参数。
如果回调函数的返回值不是 None，这个返回值将会代替默认的 response 用作新的 HttpResponse 对象。

```
SimpleTemplateResponse.render()
```

将 response.content 对象赋值给 SimpleTemplateResponse.rendered_content，执行全部回调函数，返回新的 response 对象。
只有第一次调用 render() 方法时才生效，后续调用只会返回第一次生成的 response 对象。

9.8.2 TemplateResponse 对象

TemplateResponse 是 SimpleTemplateResponse 的子类，构造方法定义如下：

```
TemplateResponse.__init__(request, template, context=None, content_type=None, status=None, charset=None, using=None)
```

各参数介绍如下。
- request：HttpRequest 对象实例。
- 其他参数与 SimpleTemplateResponse.__init__() 的参数含义一样。

9.8.3 TemplateResponse 对象渲染过程

TemplateResponse 对象实例在发送给客户浏览器之前必须完成渲染工作。以下三种情形 TemplateResponse 对象会被渲染：

- ❏ 明确调用 TemplateResponse.render() 方法的时候。
- ❏ 显式使用 response.content 为响应对象赋值的时候。
- ❏ 通过 template response 中间件，但是还没有通过 response 中间件。

注意

TemplateResponse 对象只能够被渲染一次，渲染完成后，继续调用 render() 方法将不起任何作用。但是，如果重新为 response.content 赋值，文档内容的改变还是会更新的。

下面介绍几种渲染 TemplateResponse 对象的方式。在演示之前先准备两个模板文件。

模板文件 1：original.html。

```
<!DOCTYPE HTML>
<html>
<head>
    <meta http-equiv="Content-Type" content="text/html; charset=gb2312;" />
    <meta http-equiv="Content-Language" content="zh-cn" />
</head>

<body>
    <div id="content">
        {{say_hi}}
    </div>
</body>
</html>
```

模板文件 2：other.html。

作为参照，这个模板非常简单，只需要输出 {{say_hi}} 即可。

下面使用不同方法对 TemplateResponse 进行渲染。

方法 1：初始化 TemplateResponse 对象并使用 render 方法进行渲染，如下图所示。

```
>>> from django.template.response import TemplateResponse
>>> from django.http import HttpRequest
>>> t = TemplateResponse(HttpRequest(), 'D:/Code/django3/mysite/blog/templates/original.html', {'say_hi': 'Hello Django!'})
>>> t.render()
<TemplateResponse status_code=200, "text/html; charset=utf-8">
>>> print(t.content)
b'<!DOCTYPE HTML>\n<html>\n<head>\n    <meta http-equiv="Content-Type" content="text/html; charset=gb2312;" />\n    <meta http-equiv="Content-Language" content="zh-cn" />\n</head>\n\n<body>\n    <div id="content">\n        Hello Django!\n    </div>\n</body>\n</html>\n'
>>>
```

方法 2：变更模板文件重新渲染，如下图所示。

```
>>> t.template_name = 'D:/Code/django3/mysite/blog/templates/other.html'
>>> t.render()
<TemplateResponse status_code=200, "text/html; charset=utf-8">
>>> print(t.content)
b'<!DOCTYPE HTML>\n<html>\n<head>\n    <meta http-equiv="Content-Type" content="text/html; charset=gb2312;" />\n    <meta http-equiv="Content-Language" content="zh-cn" />\n</head>\n\n<body>\n    <div id="content">\n        Hello Django!\n    </div>\n</body>\n</html>\n'
>>>
```

方法 3：直接修改 response.content 属性，如下图所示。

```
>>> t.content = t.rendered_content
>>> print(t.content)
b'Hello Django!'
>>>
```

9.8.4 回调函数

某些操作必须基于一个完全渲染结束的 response 对象，如缓存操作。如果使用中间件来处理这些操作，一切都能正常执行，因为中间件保证所有操作都是在渲染结束之后才可以执行。但是，如果使用视图装饰器，就会出问题，因为装饰器是被立即执行的。为了解决这些问题，TemplateResponse 允许开发人员注册回调函数，TemplateResponse 的回调函数是在模板渲染结束后被立即执行的。下面是一个回调函数的示例代码：

```
from django.template.response import TemplateResponse

def my_render_callback(response):
    # 完成回调任务
    do_post_processing()

def my_view(request):
    # 创建 TemplateResponse 对象
    response = TemplateResponse(request, 'mytemplate.html', {})
    # 注册回调函数
    response.add_post_render_callback(my_render_callback)
    # 返回 response 对象
    return response
```

9.8.5 使用 TemplateResponse 对象

TemplateResponse 对象可以像 HttpResponse 对象一样使用，没有任何限制。
下面修改 blog/index 视图，使用 TemplateResponse 对象替代 HttpResponse：

```
def Index(request):
    blogs = Blog.objects.all()
    context = {
        "blogs": blogs
    }
    return TemplateResponse(request, "index.html", context)
```

打开浏览器，效果如下图所示。

9.9 文件上传

文件上传是一个比较常用的网站功能，在服务器端，Django 会使用一个叫作 request.FILES 的对象来处理上传的文件。本节主要介绍 Django 是如何保存文件的。

9.9.1 单一文件上传

单一文件上传的步骤如下。

步骤 1：新建文件 forms.py，编写表单类，该表单包含一个 FileField 字段。

```
from django import forms

class UploadFileForm(forms.Form):
    file = forms.FileField()
```

步骤 2：下面代码用于处理 UploadFileForm 表单的视图，文件名是 form_view.py。

```
import os
import sys
from django.http import HttpResponseRedirect
from django.shortcuts import render, reverse
from .forms import UploadFileForm
from django.conf import settings

def upload_file(request):
    if request.method == 'POST':
        form = UploadFileForm(request.POST, request.FILES)
        if form.is_valid():
            f = request.FILES['file']
            handle_uploaded_file(f)
            return HttpResponseRedirect(reverse('blog:success', args=(f.name,)))
    else:
        form = UploadFileForm()
        return render(request, 'upload.html', {'form': form})

def handle_uploaded_file(f):
    p = os.path.join(settings.MEDIA_ROOT, 'upload', f.name)
    with open(p, 'wb+') as destination:
        for chunk in f.chunks():
            destination.write(chunk)

def success(request, name):
    return render(request, 'success.html', {'file': name})
```

步骤 3：创建模板文件 upload.html。

```
<form action="{% url 'blog:upload' %}" method="post" enctype="multipart/form-data">
    {% csrf_token %}
    {{ form }}
    <input type="submit" value="Submit" />
</form>
```

步骤 4：创建模板文件 success.html，文件内容如下：

```
{{ file }} 上传成功！
```

步骤 5：添加 URL 路由。

```
from django.urls import path
from . import views, form_view
app_name = 'blog'

urlpatterns = [
    ...
    path(r'upload/', form_view.upload_file, name='upload'),
    path(r'success/<str:name>/', form_view.success, name='success'),
]
```

代码分析：
- 只有 POST 请求方式才会触发文件上传动作。
- request.FILES 是一个字典对象，包含所有上传文件，字典的 Key 是表单类的字段，本例中的 Key 是 "file"。
- 用于文件上传操作的表单元素需要包含 enctype="multipart/form-data" 属性。
- 为了避免 read() 方法一次性将文件读取到内存中造成内存不足的问题，使用 f.chunks() 方式将文件分块处理。
- 文件上传到 settings.MEDIA_ROOT 所指定路径下的 upload 文件夹中，注意该路径须提前创建好。

代码演示步骤如下。

步骤 1：打开 http://127.0.0.1:8000/blog/upload/，如下图所示。

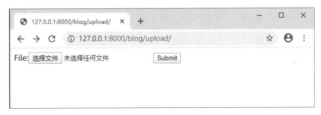

查看网页源代码，可以看到 form 表单包含一个叫作 enctype 的属性以及一个隐藏的 csrf 元素，如下图所示。

```
1  <form action="/blog/upload/" method="post" enctype="multipart/form-data">
2      <input type="hidden" name="csrfmiddlewaretoken" value="AqGlsoIpprchBWwPDCDBcRIqPn22E1WCk7puPU8cNrdYuwUv03cg1W7F0Rh6lSmN">
3      <tr><th><label for="id_file">File:</label></th><td><input type="file" name="file" required id="id_file"></td></tr>
4      <input type="submit" value="Submit" />
5  </form>
```

步骤 2：选择本地文件，如下图所示。

步骤 3：单击"Submit"按钮上传文件，如下图所示。

上传成功后，可以看到 settings.MEDIA_ROOT 路径下已经存在对应的文件了。

9.9.2 多文件上传

由于标准的 HTML 只允许使用 <input type="file"> 进行文件上传，而 <input type="file"> 每次只能上传一个文件，因此对于需要进行大量文件上传的操作来说会很不方便，这在 Django 中就变得相对简单很多。

具体步骤如下。

步骤 1：添加 FileFieldForm 表单类。

```
class FileFieldForm(forms.Form):
    file_field = forms.FileField(widget=forms.ClearableFileInput(attrs={'multiple': True}))
```

步骤 2：添加用于处理多个文件上传的视图方法。

```
from django.views.generic.edit import FormView
from .forms import FileFieldForm

class FileFieldView(FormView):
    form_class = FileFieldForm
    template_name = 'multyupload.html'
    success_url = u'/blog/success/ 全部文件'

    def post(self, request, *args, **kwargs):
        form_class = self.get_form_class()
        form = self.get_form(form_class)
        files = request.FILES.getlist('file_field')
        if form.is_valid():
            for f in files:
                handle_uploaded_file(f)
            return self.form_valid(form)
        else:
            return self.form_invalid(form)
```

步骤 3：创建模板文件 multyupload.html。

```
<form action="{% url 'blog:multyupload' %}" method= "post" enctype= "multipart/form-data" >
    {% csrf_token %}
    {{ form }}
    <input type= "submit" value= "Submit" />
</form>
```

步骤 4：添加 URL 路由。

```
path(r'multyupload/', form_view.FileFieldView.as_view(), name='multyupload'),
```

此时每个 HTTP 请求都可以上传多个文件，浏览器访问效果如下图所示。

将鼠标光标移动到"5 个文件"上，可以看到具体上传的文件列表，效果如下图所示。

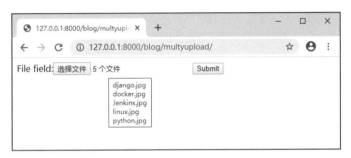

9.9.3 临时文件

从文件开始上传到最终保存到磁盘上，这中间需要经过一个过渡。在这个过程中，如果文件小于 2.5MB，那么 Django 会将文件存储在内存中；如果文件大于 2.5MB，Django 就会在系统的临时目录中生成一个临时文件，例如 Linux 系统中就会生成类似 /tmp/tmpzfp6I6.upload 的文件。如果上传的文件足够大，你还能观察到这个临时文件一直在增长。

9.10 类视图

除了前面介绍的视图方法外，Django 还提供了一系列类视图，通过使用类视图可以提高代码重用率。Django 一共提供了几十个类视图，详细信息可以参考 Django 官方文档：https://docs.djangoproject.com/zh-hans/3.0/topics/class-based-views/。

本节将对 Django 的类视图进行简单概述，并选择部分常用视图进行讲解。

9.10.1 类视图入门

Django 提供了一些基本的视图类，这些类可以直接拿来使用。类视图的英文名字叫作 Class-Based View，对应的函数视图叫作 Function-Based View。下面用 TemplateView 来演示如何使用 Django 的类视图。

首先修改 URL：

```
from django.urls import path
from django.views.generic import TemplateView

urlpatterns = [
    path(r'about/', TemplateView.as_view(template_name="about.html")),
]
```

然后编写一个模板文件"about.html"，该模板只包含一句话。

浏览器访问效果如下图所示。

9.10.2 继承类视图

如果 Django 提供的类视图不能满足工作需要，还可以基于已有的类视图开发新的视图。

对 TemplateView 进行重写，代码如下：

```
from django.views.generic import TemplateView

class AboutView(TemplateView):
    template_name = "about.html"
```

添加 URL：

```
path(r'about/', views.AboutView.as_view()),
```

此时重新访问 http://127.0.0.1:8000/blog/about/，显示效果一样。

开发新的类视图的好处是，可以高度定制化类视图，如修改类属性、修改数据方法等。

9.11 通用视图

很多时候我们都是通过视图显示一个列表或者一个对象的详细信息。正是由于类似的工作比较常见，因此 Django 提供了一些通用类视图。

9.11.1 通用视图

下面仍然以 Blog 应用程序为例，展示如何使用通用类视图。

步骤1：将 index 视图方法改写为视图类。

```
from django.views.generic import ListView
from .models import Blog

class index(ListView):
    model = Blog
```

步骤 2：修改 URL。

```
path('', views.index.as_view(), name='default'),
path(r'index/', views.index.as_view(), name='index'),
```

步骤 3：创建模板。

以上就是全部的 Python 代码，接下来添加模板。默认情况下，Django 会在 tempaltes 文件夹下查找一个与应用程序同名的文件夹，本例中文件夹名字是"blog"，然后在"blog"文件夹下查找名为"模型_list.html"的模板，本例中默认模板名为"blog_list.html"。

在 templates 文件夹下创建模板"blog/blog_list.html"：

```
{% extends "base.html" %}
{% load static %}

{% block content %}

    {% for blog in object_list %}
        <h2 class="blog_head">{{ blog.id }} - {{ blog.name }}</h2>
        <p class="blog_body">
            {{ blog.tagline }}
        </p>
    {% endfor %}

{% endblock %}
```

> **注意**
>
> 由于 ListView 是通用视图，因此它传递给模板的上下文对象是 object_list，而不是具体的模型名。

浏览器访问效果如下图所示。

也可以使用 template_name 属性来指定模板：

```
class index(ListView):
    model = Blog
    template_name = 'blog/templates/index.html'
```

9.11.2 修改通用视图属性

细心的读者可能发现了，在前面的模板中使用 object_list 变量保存全部博客信息，这个变量很不直观，其实 Django 还提供了一个更友好的名字"小写的模型名_list"，本例中对应的变量名为"blog_list"。因此，在不改变视图代码的情况下，模板中的 object_list 变量可以直接用 blog_list 替代：

```
{% extends "base.html" %}
{% load static %}

{% block content %}

    {% for blog in blog_list %}
        <h2 class="blog_head">{{ blog.id }} - {{ blog.name }}</h2>
        <p class="blog_body">
            {{ blog.tagline }}
        </p>
    {% endfor %}

{% endblock %}
```

如果对 Django 提供的这两种变量命名方式都不满意，还可以通过修改 context_object_name 属性来自定义对象变量：

```
class index(ListView):
    model = Blog
    context_object_name = 'blogs'
```

此时需要将原有模板中的上下文对象修改为"blogs"，在浏览器中访问时效果一样。

9.11.3 添加额外的上下文对象

有时在模板中不只需要默认的模型数据列表，可能还需要额外的一些信息，此时可以通过重写 get_context_data() 方法达到这样的目的。

下面修改代码使得博客列表中只显示最近发布的三篇博客文章，并显示全部作者。

修改视图类：

```
class index(ListView):
    model = Blog
    context_object_name = 'blogs'

    def get_context_data(self, **kwargs):
        context = super().get_context_data(**kwargs)
        context['entries'] = Entry.objects.all().order_by('pub_date')[0:3]
        context['author_list'] = Author.objects.all()
        return context
```

修改模板：

```
{% extends "base.html" %}
{% load static %}

{% block content %}

    {% for author in author_list %}
        <span style="background-color:grey; margin:10px;">
            {{ author.name }}
        </span>
    {% endfor %}

    {% for entry in entries %}
        <h2 class="blog_head">{{ entry.blog.id }} - {{ entry.blog.name }}</h2>
        <p class="blog_body">
            {{ entry.blog.tagline }}
        </p>
    {% endfor %}

{% endblock %}
```

新模板在浏览器中的显示效果如下图所示。

9.11.4　queryset 属性

前面使用"model = Blog"在通用视图中指定模型，其实这句代码是 queryset 的简写形式，标准写法是：

```
queryset = Blog.objects.all()
```

通过使用 queryset 可以对数据进行一定的修改，下面重新编写 index 视图：

```
class index(ListView):
    queryset = Blog.objects.all()[:3]
    context_object_name = 'blogs'

    def get_context_data(self, **kwargs):
```

```
        context = super().get_context_data(**kwargs)
        context['entries'] = Entry.objects.all().order_by('pub_date')[0:3]
        context['author_list'] = Author.objects.all()
        return context
```

此时网页仍然显示前三篇博客文章。

9.11.5 动态过滤

通用视图可以通过传递 URL 参数的方式过滤显示内容，修改 index 的 URL，使得 index 页面允许接收 Author 作为过滤条件：

```
path(r'index/<author>/', views.index.as_view(), name='index'),
```

修改 index 视图：

```
class index(ListView):
    model = Entry
    template_name = 'blog/blog_list.html'
    context_object_name = 'entries'

    def get_queryset(self):
        self.author = get_object_or_404(Author, name=self.kwargs['author'])
        return Entry.objects.filter(authors=self.author)

    def get_context_data(self, **kwargs):
        context = super().get_context_data(**kwargs)
        context['entries'] = Entry.objects.filter(authors__id__contains=self.author.id).order_by('pub_date')[0:3]
        context['author_list'] = Author.objects.all()
        return context
```

浏览器显示效果如下图所示。

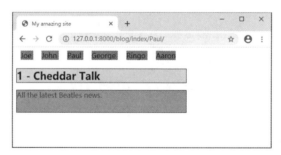

当搜索的作者没有发表任何博客时，网页不显示任何博客文章，如下图所示。

当搜索的作者不存在时，网页提示没有匹配到任何作者，如下图所示。

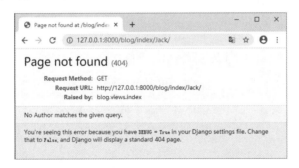

9.11.6 通用视图与模型

从前面的示例代码不难发现，某些视图指定了 model 属性，而某些视图并没有给出 model 属性，但是所有这些视图都能够很好地工作，这是由于通用视图可以从以下三个方面确定模型：
- 视图的 model 属性指定模型。
- 视图的 get_object() 方法返回模型。
- 视图的 queryset 所使用的模型。

9.12 表单视图

Django 的表单视图可以处理一些基本的表单任务，如表单验证、编辑等。以前面的 FileFieldView 视图为例，如果用户没有选择任何文件时就单击提交按钮，表单会阻止 HTTP 请求并弹出提示信息，如下图所示。

另外，当用户提交合法信息后，还可以使用 form_valid() 方法执行额外操作，下面修改 FileFieldView 视图，添加 form_valid() 方法：

```
class FileFieldView(FormView):
    ...

    def form_valid(self, form):
        print('data: ', form.data)
        return super().form_valid(form)

    def post(self, request, *args, **kwargs):
...
        if form.is_valid():
            for f in files:
                print(type(f))  # 打印文件信息
```

```
                    handle_uploaded_file(f)
            return self.form_valid(form)
        else:
            return self.form_invalid(form)
```

重启 Web 服务并上传文件，此时服务器输出以下信息：

```
[19/Apr/2020 17:37:23] "GET /blog/multyupload/ HTTP/1.1" 200 408
<class 'django.core.files.uploadedfile.InMemoryUploadedFile'>
<class 'django.core.files.uploadedfile.InMemoryUploadedFile'>
data: <QueryDict: {'csrfmiddlewaretoken': ['qA8BZf53zkkzxzbBT62piip0jNtK8PlDgwOx5i
TeShoqiu1AQwiHCk7XowQlLtGR']}>
[19/Apr/2020 17:37:34] "POST /blog/multyupload/ HTTP/1.1" 302 0
[19/Apr/2020 17:37:34] "GET /blog/success/%E5%85%A8%E9%83%A8%E6%96%87%E4%BB%B6 
HTTP/1.1" 301 0
[19/Apr/2020 17:37:34] "GET /blog/success/%E5%85%A8%E9%83%A8%E6%96%87%E4%BB%B6/ 
HTTP/1.1" 200 27
```

9.12.1 编辑表单视图

通过使用表单视图，可以在编写很少代码的情况下完成模型的增、删、改操作。下面以 Author 模型为例介绍如何使用表单视图编辑模型。

步骤 1：修改 Author 模型类。

```python
from django.db import models
from django.shortcuts import reverse

class Author(models.Model):
    name = models.CharField(max_length=200)
    email = models.EmailField()

    def __str__(self):
        return self.name

    def get_absolute_url(self):
        return reverse('blog:author-detail', kwargs={'pk': self.pk})
```

步骤 2：添加视图。

```python
from django.views.generic import DetailView
from django.views.generic.edit import CreateView, UpdateView, DeleteView
from django.urls import reverse_lazy
from .models import Author

class AuthorDetail(DetailView):
    queryset = Author.objects.all()
    template_name = 'blog/author/author_detail.html'

class AuthorCreate(CreateView):
    model = Author
    fields = ['name', 'email']
    template_name = 'blog/author/author_form.html'

class AuthorUpdate(UpdateView):
```

```python
    model = Author
    fields = ['name', 'email']
    template_name = 'blog/author/author_update.html'

class AuthorDelete(DeleteView):
    model = Author
    success_url = reverse_lazy('blog:success', args=['删除成功'])
    template_name = 'blog/author/author_confirm_delete.html'
```

步骤 3：添加模板 blog\templates\blog\author\author_detail.html。

```
{% extends "base.html" %}
{% load static %}

{% block content %}

    <div>
        <div>{{ object.name }}</div>
        <div>{{ object.email }}</div>
    </div>

{% endblock %}
```

步骤 4：添加模板 blog\templates\blog\author\author_form.html。

```html
<form action="{% url 'blog:author-add' %}" method="post">
    {% csrf_token %}
    {{ form }}
    <input type="submit" value="Submit" />
</form>
```

步骤 5：添加模板 blog\templates\blog\author\author_update.html。

```html
<form action= "{% url 'blog:author-update' object.id %}" method= "post" >{% csrf_token %}
    {{ form.as_p }}
    <input type="submit" value="Update" />
</form>
```

步骤 6：添加模板 blog\templates\blog\author\author_confirm_delete.html。

```html
<form action= "{% url 'blog:author-delete' object.id %}" method= "post" >{% csrf_token %}
    <p>确定要删除 "{{ object.name }}" 吗？ </p>
    <input type="submit" value="Confirm" />
</form>
```

步骤 7：添加 URL。

```python
from . import author_view

urlpatterns = [
    ...
    path(r'author/<int:pk>/', author_view.AuthorDetail.as_view(), name='author-detail'),
    path(r'author/add/', author_view.AuthorCreate.as_view(), name='author-add'),
    path(r'author/<int:pk>/update/', author_view.AuthorUpdate.as_view(), name='author-update'),
```

```
        path(r'author/<int:pk>/delete/', author_view.AuthorDelete.as_view(), name='author-
delete'),
]
```

需要注意的是，UpdateView 和 DeleteView 对于 POST 和 GET 请求都有不同的处理方式，GET 请求只会用于显示确认信息，而 POST 请求才会真正执行操作。例如更新 Author 时首先弹出确认框，只有单击"Update"按钮之后才会真正更新信息，如下图所示。

Delete 操作同样需要先确认才可以真正执行，如下图所示。

9.12.2 当前用户

一般进行数据操作时都需要记录执行人，也就是当前登录的用户，Django 使用 User 模块进行用户管理，下面修改 Author 模型使其能够记录添加作者操作的用户信息：

```
from django.contrib.auth.models import User

class Author(models.Model):
    ...
    created_by = models.ForeignKey(User, null=True, on_delete=models.CASCADE)
    ...
```

修改 AuthorCreate 视图：

```
class AuthorCreate(CreateView):
    model = Author
    fields = ['name', 'email']

    def form_valid(self, form):
        form.instance.created_by = self.request.user
        return super().form_valid(form)
```

在 blog/admin.py 中添加以下代码：

```
from .models import *
admin.site.register(Author)
```

执行 migrations 命令。

重新启动 Web 服务并登录到管理后台，此时可以添加作者，效果如下图所示。

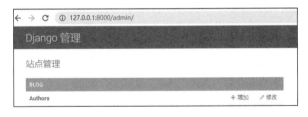

添加完成后查看数据库，效果如下图所示。

这里的 created_by_id=1 的用户就是系统超级用户（位于 auth_user 表）。

第 10 章
模板

模板可以看作创建 HTML 页面的样本。模板包含静态的 HTML 和用于描述如何动态生成 HTML 的特殊语法两个部分。模板的结构和 HTML 文件非常相似，甚至完全可以使用一个 HTML 文件作为模板。

Django 使用模板引擎对模板文件进行解释，一个 Django 工程可以配置一个或者多个模板引擎。如果项目中没有使用模板，那么也可以不配置模板引擎。Django 自带的模板系统叫作 Django Template Language（DTL），通过该引擎可以方便地加载模板文件并在内存中进行编译，然后插入动态数据，最后返回一个字符串。

第 5 章已经详细介绍了如何配置模板引擎，这里不再赘述。

10.1 加载模板

django.template.loader 模块中提供了两个用于加载模板的方法：
- get_template(template_name, using=None)：该方法接收一个模板名，返回 Template 对象。可以通过修改 using 参数的方式更改模板引擎。
- select_template(template_name_list, using=None)：该方法接收一个模板名称的列表，返回第一个存在的 Template 对象。同样 using 参数用于改变模板引擎。

当找不到对应的模板时，这两个方法都会返回 TemplateDoesNotExist 异常。如果模板找到了，但是模板中存在语法错误，返回 TemplateSyntaxError 异常。

对于不同的模板引擎，Template 对象也是不同的，但是 Template 对象必须包含一个 render() 方法。Render() 的语法结构如下：

```
Template.render(context=None, request=None)
```

参数说明：
- context 是需要展示到 HTML 文件上的数据集合，它是一个字典对象。如果 render 没有接收任何 context，模板引擎就会直接渲染模板而不插入任何数据。
- request 是一个 HttpRequest 对象。不同的模板引擎对 request 对象的处理方式不同。

下面看看搜索模板的逻辑。假设存在如下模板配置：

```
TEMPLATES = [
    {
        'BACKEND': 'django.template.backends.django.DjangoTemplates',
        'DIRS': [
```

```
            '/home/html/example.com',
            '/home/html/default',
        ],
    },
    {
        'BACKEND': 'django.template.backends.jinja2.Jinja2',
        'DIRS': [
            '/home/html/jinja2',
        ],
    },
]
```

下面是调用 get_template('story_detail.html') 时查找模板的顺序：

（1）/home/html/example.com/story_detail.html ('django' engine)

（2）/home/html/default/story_detail.html ('django' engine)

（3）/home/html/jinja2/story_detail.html ('jinja2' engine)

下面是调用 select_template(['story_253_detail.html', 'story_detail.html']) 时查找模板的顺序：

（1）/home/html/example.com/story_253_detail.html ('django' engine)

（2）/home/html/default/story_253_detail.html ('django' engine)

（3）/home/html/jinja2/story_253_detail.html ('jinja2' engine)

（4）/home/html/example.com/story_detail.html ('django' engine)

（5）/home/html/default/story_detail.html ('django' engine)

（6）/home/html/jinja2/story_detail.html ('jinja2' engine)

一旦 Django 找到匹配的模板，查询工作停止，即使后面存在其他匹配的模板文件也不会继续查询。这样做可以通过子文件夹分别管理不同应用程序的模板文件，使整个工程结构变得清晰。

如果使用文件路径查找模板，例如：

```
get_template('news/story_detail.html')
```

对于同样的模板配置，将会在下面的路径进行查找：

（1）/home/html/example.com/news/story_detail.html ('django' engine)

（2）/home/html/default/news/story_detail.html ('django' engine)

（3）/home/html/jinja2/news/story_detail.html ('jinja2' engine)

10.2 模板语言

Django 的模板语言非常强大，有 HTML 开发经验的人对此会感到非常亲切。本节将详细介绍 Django 的模板语言。

Django 的模板是一个简单的文本文件，可以是任何文本格式，如 HTML、XML、TXT 等，推荐使用 .HTML 格式。模板语言主要包括变量（Variables）、标签（Tags）、过滤器（Filters）和注释（Comments）四部分。

10.2.1 变量

在模板中变量是使用 {{ 和 }} 包围起来的对象，它的值存放在上下文对象（context）中，上下文对象中可能存在很多变量，这些变量以类似字典（dict）的形式存放。

变量名可以包括字母、数字和下画线,但是绝对不可以包括空格和其他标点符号。

英文句点(.)在变量中有特殊意义,如果模板引擎遇到了句点,将会按照下面的顺序对其进行解释。

- 字典查找:{{ my_dict.Key }}。
- 查找属性和方法:{{ my_object.attribute }}。
- 查找下标:{{ my_list.0 }}。

> **注意**
>
> 如果句点后面的变量是一个方法,那么这个方法会按照空参数的方式调用。例如一个字典的 iteritems 方法可以在模板中用以下方式调用:
>
> ```
> {% for k, v in defaultdict.items %}
>     Do something with k and v here...
> {% endfor %}
> ```

10.2.2 标签

标签的用法类似于 {% tag %}。相对于变量来说标签更加复杂,标签可以用于输出文本、控制代码执行逻辑等。

有些标签还需要有开始标记和结束标记,这类标签的格式类似于 {% tag %}…tag contents…{% endtag %}。

Django 内置了 20 多种标签,具体介绍可以参看 Django 官方文档:

https://docs.djangoproject.com/zh-hans/3.0/ref/templates/builtins/#ref-templates-builtins-tags

下面选择部分常用标签进行介绍。

1. block

用于定义一个模板块,这个模板块可以被子模板重写,block 的用法如下:

```
{% block 模板块的名字 %}
    ...
{% endblock %}
```

10.4 节将会详细介绍 block 标签。

2. comment

模板中的注释,模板引擎将会忽略 {% comment %} 和 {% endcomment %} 之间的任何代码。

comment 的用法如下:

```
{% comment "Optional note" %}
    <p>Commented out text with {{ create_date|date:"c" }}</p>
{% endcomment %}
```

3. cycle

循环提取参数值,当遍历到最后一个参数时,如果外部循环仍然没有停止,则从第一个参数重新开始循环。例如下面代码:

```
{% for o in some_list %}
    <tr class="{% cycle 'row1' 'row2' %}">
        ...
    </tr>
{% endfor %}
```

假设 some_list 的长度大于 2，那么输出的第一个 <tr> 标签的 class 是 row1，第二个 <tr> 标签的 class 是 row2，第三个 <tr> 标签的 class 重新使用 row1，直到循环结束。

4. extends

用于标记当前模板继承自哪个父模板。

extends 标签有以下两种使用形式：

- {% extends "base.html" %}
- {% extends variable %}

10.4 节将会详细介绍 extends 标签。

5. for

循环遍历一个列表。

例如下面代码使用 for 循环生成一个无序列表：

```
<ul>
{% for athlete in athlete_list %}
    <li>{{ athlete.name }}</li>
{% endfor %}
</ul>
```

还可以使用 reversed 对列表进行翻转：

```
{% for obj in list reversed %}
{% endfor %}
```

使用 for 循环遍历字典：

```
{% for Key, value in data.items %}
    {{ Key }}: {{ value }}
{% endfor %}
```

for 循环提供了一些变量，如下表所示。

变量	说明
forloop.counter	当前循环位置（以数字 1 位起始）
forloop.counter0	当前循环位置（以数字 0 位起始）
forloop.revcounter	反向循环位置（列表的最后一位是 1，列表第一位是 n）
forloop.revcounter0	反向循环位置（列表的最后一位是 0，列表第一位是 n-1）
forloop.first	如果是当前循环的第一位，则返回 True
forloop.last	如果是当前循环的最后一位，则返回 True
forloop.parentloop	上级循环

接下来通过一个示例进行介绍。

视图代码：

```
class AboutView(TemplateView):
    template_name = "about.html"

    def get_context_data(self, **kwargs):
        context = super().get_context_data(**kwargs)
        context['numbers'] = [0, 1,2,3]
        return context
```

模板代码：

```
{% for n in numbers %}
    {% if forloop.first %}
        这是循环的起始位置 {{ n }}<br />
    {% endif %}

    {{ forloop.counter }} - {{ n }}<br />

    {% if forloop.last %}
        这是循环的结束位置 {{ n }}<br />
    {% endif %}
{% endfor %}
```

浏览器显示效果如下图所示。

代码示例：

```
{% for n in numbers %}
   {% for n in numbers %}
        {{ forloop.parentloop.counter0 }} - {{ n }}<br />
   {% endfor %}
{% endfor %}
```

浏览器显示效果如下图所示。

6. for … empty

当被遍历对象为空时，显示 empty 标签内容，其他部分与 for 循环一致。

代码示例：

```
<ul>
{% for athlete in athlete_list %}
    <li>{{ athlete.name }}</li>
{% empty %}
    <li>Sorry, no athletes in this list.</li>
{% endfor %}
</ul>
```

7. if

条件判断标签，当判断条件为真时（存在、非空、非 False 值）输出标签内容。与 Python 一样，if 标签也支持 elif 和 else 条件分支语句。代码示例：

```
{% if athlete_list %}
    Number of athletes: {{ athlete_list|length }}
{% elif athlete_in_locker_room_list %}
    Athletes should be out of the locker room soon!
{% else %}
    No athletes.
{% endif %}
```

同样 if 标签还支持对判断条件进行逻辑运算，逻辑运算符包括 and、or、not：

```
{% if athlete_list and coach_list %}
    Both athletes and coaches are available.
{% endif %}
```

> **注意**
>
> 不能在 if 标签中使用圆括号对判断条件进行分组，如以下代码将会抛出"Could not parse the remainder: '(False' from '(False'"错误：
>
> ```
> {% if numbers and (False or True) %}
>     Pass
> {% endif %}
> ```

除了逻辑运算符外，if 标签还支持的运算符包括：==、!=、<、>、<=、>=、in、not in、is、is not。代码示例：

```
{% if somevar == "x" %}
  This appears if variable somevar equals the string "x"
{% endif %}
```

最后，if 语句还可以使用过滤器，如使用 length 过滤器显示长度大于等于 100 个字符的信息：

```
{% if messages|length >= 100 %}
   You have lots of messages today!
{% endif %}
```

8. include

加载其他模板，例如加载 "foo/bar.html"：

```
{% include "foo/bar.html" %}
```

如果新模板中包含变量，include 标签还可以传递变量值，例如模板"name_snippet.html"形式如下：

```
{{ greeting }}, {{ person|default:"friend" }}!
```

可以使用 include 在引用"name_snippet.html"的同时传递参数：

```
{% include "name_snippet.html" with person="Jane" greeting="Hello" %}
```

如果在引用"name_snippet.html"的时候只传递一个变量值而忽略其他变量，可以使用 only 标签：

```
{% include "name_snippet.html" with greeting="Hi" only %}
```

9. load

加载自定义模板标签。

例如加载 somelibrary 和 package.otherlibrary 中的全部标签：

```
{% load somelibrary package.otherlibrary %}
```

加载 somelibrary 中的 foo 和 bar 两个标签：

```
{% load foo bar from somelibrary %}
```

10. now

显示当前日期或时间。

例如：

```
{% now "jS F Y H:i" %}
```

11. url

动态生成 url，例如：

```
{% url 'some-url-name' v1 v2 %}
{% url 'some-url-name' arg1=v1 arg2=v2 %}
```

详细使用方法可参见 7.8 节。

12. with

为复杂变量创建别名，尤其是使用句点访问多级变量时非常方便：

```
{% with total=business.employees.count %}
    {{ total }} employee{{ total|pluralize }}
{% endwith %}
```

10.2.3 人性化语义标签

除了上述功能性标签外，Django 还提供了很多辅助性标签，这些标签只是为了使变量输出变得更加可读，下面对这些标签进行简单介绍。

首先为了使用这些标签，需要在 INSTALLED_APPS 中注册"django.contrib.humanize"，然后在

模板中引用 {% load humanize %} 就可以了。

下面是具体标签介绍。

1. apnumber

将数字 1～9 转换为对应的单词，但是其他数字不转换，如数字 10 将被原样输出。

示例：

数字 1 被转换为 one。

数字 2 被转换为 two。

数字 10 仍显示 10。

如果当前工程语言是中文，数字将会被转换为对应的汉字，例如：

```
{{ 1|apnumber }}
{{ 2|apnumber }}
{{ 5|apnumber }}
{{ 10|apnumber }}
```

输出结果如下图所示。

2. intcomma

输出以逗号分隔的数字，例如：

```
{{ 4500|intcomma }}
{{ 4500.2|intcomma }}
```

输出结果如下图所示。

需要注意的是，如果当前语言不支持以逗号分隔数值类型，那么 intcomma 将不生效，如当 LANGUAGE_CODE = 'zh-hans' 时数值类型将原样显示。

3. intword

以可读性较高的文字形式输出超大的数字，如 1000000 输出 "1.0 million"，1200000 输出 "1.2 Million"。在 Dango 3.0 中除了 1.0 外，其他输出字符串都以复数形式表示。

对于中文系统，将会输出对应的中文，如 1200000 输出 "1.2 百万"。

代码示例：

```
{{ 1000000|intword }}
{{ 1000000.1|intword }}
{{ 1200000000|intword }}
```

英文环境下的显示效果如下图所示。

中文环境下的显示效果如下图所示。

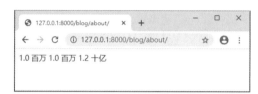

4. naturalday

将当前日期以及前后一天分别输出为"today""yesterday"和"tomorrow",而中文系统分别输出"今天""昨天"和"明天"。其他日期使用格式化字符串进行转换显示。

修改 about 视图,增加 'now'、'yesterday'、'tomorrow' 和 'other_day' 四个上下文变量:

```
class AboutView(TemplateView):
    template_name = "about.html"

    def get_context_data(self, **kwargs):
        context = super().get_context_data(**kwargs)

        yesterday_timedelta = datetime.timedelta(days=-1)
        tomorrow_timedelta = datetime.timedelta(days=1)
        now = datetime.datetime.now()
        yesterday = now.__add__(yesterday_timedelta)
        tomorrow = now.__add__(tomorrow_timedelta)

        context['now'] = now
        context['yesterday'] = yesterday
        context['tomorrow'] = tomorrow
        context['other_day'] = datetime.datetime(year=2010, month=1, day=5)
        return context
```

在 about 模板中增加对以上变量的显示:

```
{{ yesterday|naturalday }}
{{ now|naturalday }}
{{ tomorrow|naturalday }}
{{ other_day|naturalday }}
```

输出结果如下图所示。

5. naturaltime

对于日期时间（datetime）格式，将变量所表示的时间值与系统当前时间进行比较，如时间变量早于当前时间则输出"XX 秒/分钟/小时前"，如时间变量晚于当前时间则输出"XX 秒/分钟/小时以后"。

仍以 'now'、'yesterday'、'tomorrow' 和 'other_day' 为例查看显示效果。

修改模板代码：

```
{{ yesterday|naturaltime }}
{{ now|naturaltime }}
{{ tomorrow|naturaltime }}
{{ other_day|naturaltime }}
```

输出结果如下图所示。

6. ordinal

将数字转换为序数，如 1 输出"1st"，2 输出"2nd"，3 输出"3rd"。

10.2.4 过滤器

过滤器可以用来修改变量的显示样式。

过滤器的使用方式 {{ 变量|过滤器方法 }}。过滤器可以连续使用，形式如：{{ 变量|过滤器方法 1|过滤器方法 2}}。

> **注意**
>
> 变量、管道符（|）和过滤器方法之间不能有空格。

某些过滤器还可以接收参数，例如 {{bio|truncatewords:30}}，这句代码的意思是显示 bio 的前 30 个单词。

如果过滤器参数包含空格，参数就要用引号扩起来，例如 {{ list|join:", " }}。

Django 大约提供了 60 个过滤器，具体介绍可以参看 Django 官方文档：https://docs.djangoproject.com/zh-hans/3.0/ref/templates/builtins/。

下面介绍几种常用的过滤器。

1. add

加法运算：{{ value|add:"2" }}

这个方法会先按照数值来计算，如果失败了就直接将两个值拼接在一起，如链接两个数组。

如果 value 是 4 则输出 6，如果 value 是 "Django" 则输出 "Django2"；如果 value 是列表，则会进行列表拼接。

2. capfirst
首字母大写：{{ value|capfirst }}

3. center
使用指定宽度将值居中显示，例如 {{ value|center:"15" }}，如果 value 是 "Django" 则输出 "   Django   "。

4. cut
删除指定值，例如去掉字符串中的空格 {{ value|cut:" " }}，如果 value 是 "String with spaces"，那么输出 "Stringwithspaces"。

5. date
格式化日期，例如 {{ value|date:"D d M Y" }}，更多的日期格式化字符请参考附录二。

6. default
如果变量是 false 或者空，显示默认值。例如 {{ value|default:"nothing" }}，当 value=false 时，在页面上显示 nothing。

7. escape
将字符串进行 html 转义，例如：

```
{% autoescape off %}
    {{ title|escape }}
{% endautoescape %}
```

如果 title 的值是 " <Django>"，则输出 <Django>。

8. filesizeformat
将文件大小按照人类可读的形式显示，例如，一个文件有 123456789 个字节，那么使用 filesizeformat 将会显示成 117.7 MB，语法形式：{{ value|filesizeformat }}。

9. Join
拼接多个元素，例如 {{ value|join:" // " }}。

10. length
显示一个字符串或者数组的长度，例如 {{ value|length }}。

11. linenumbers
在文本前显示编号。例如 {{ value|linenumbers }}。

如果 value 是：

One

Two

Three

那么输出：

1. One

2. Two

3. Three

12. truncatewords

如果文字太长则缩短显示内容，例如 {{ value|truncatewords:2 }}，如果此时 value 是"Joel is a slug"则输出"Joel is ..."。

13. upper

字母转换大写显示，例如 {{ value|upper }}，如果 value 是"Django"则输出"DJANGO"。

10.2.5 注释

单行注释：

```
{# this won't be rendered #}
```

多行注释：

```
{% comment "Optional note" %}
    <p>Commented out text with {{ create_date|date:"c" }}</p>
{% endcomment %}
```

> 多行注释的 comment 标签不能嵌套使用。

10.3 自定义标签和过滤器

虽然 Django 提供了非常丰富的模板标签和过滤器，但是在实际工作中还是会遇到特殊需求是已有标签和过滤器所不能实现的情况，此时可以开发自定义模板标签和过滤器来实现特殊需求。

通常把自定义标签和过滤器放在 Django 应用程序的 templatetags 文件夹中，如果应用程序中没有 templatetags 文件夹，那么可以手工创建一个，这个文件夹与 models.py、views.py 文件平级，最后不要忘记在这个文件夹里面放一个 __init__.py 文件。对于手工创建的 templatetags 文件夹，必须重新启动 server 才能生效。

> templatetags 中的文件名就是将来新标签或者过滤器的名字，因此命名要谨慎，不要和其他标签、过滤器重名。

例如，下面是为应用程序 polls 添加自定义标签（poll_extras）后的目录结构：

```
polls/
    __init__.py
    models.py
    templatetags/
        __init__.py
        poll_extras.py
    views.py
```

此时可以在模板中使用下面代码加载标签 poll_extras：

```
{% load poll_extras %}
```

为了使 Python 模块成为 Django 自定义标签或过滤器，每个模块都需要包含一个名为 register 的模块级变量：

```
from django import template

register = template.Library()
```

10.3.1 编写自定义过滤器

自定义过滤器就是一个可以接收一个或者多个参数的 Python 方法：
- 接收的变量可以是任意类型，并不局限于字符串。
- 过滤器方法的参数可以有默认值。

例如，在过滤器 `{{ var|foo:"bar" }}` 中，foo 同时接收变量 var 和参数 bar。

由于模板语言不能处理异常，因此在过滤器方法中出现的异常都会成为服务器异常，应该避免在过滤器方法中出现异常情况。

下面是一个自定义过滤器的代码示例：

```
def cut(value, arg):
    """从字符串 value 中删除所有 arg"""
    return value.replace(arg, '')
```

下面是过滤器 cut 的使用方法：

```
{{ somevariable|cut:"0" }}
```

大多数过滤器并不接收参数，这类过滤器的写法类似于：

```
def lower(value): # Only one argument.
    """将字符串中的所有字母转换为小写字母"""
    return value.lower()
```

一旦编写完自定义过滤器方法，就需要使用 Library 实例对其进行注册，否则不能使用，注册代码如下：

```
register.filter('cut', cut)
register.filter('lower', lower)
```

Library.filter() 方法接收两个参数：
- 第一个参数是过滤器的名字。
- 第二个参数是过滤器方法。

除了使用 Library.filter() 方法注册过滤器外，还可以使用装饰器注册标签和过滤器，例如：

```
@register.filter(name='cut')
def cut(value, arg):
    return value.replace(arg, '')

@register.filter
def lower(value):
    return value.lower()
```

如果省略 name 参数，Django 将会使用函数名作为过滤器名字。

如果需要限制过滤器接收的变量（即 value 参数）只能是字符串，可以使用 stringfilter 方法属性，例如：

```
from django import template
from django.template.defaultfilters import stringfilter

register = template.Library()

@register.filter
@stringfilter
def blogtrans(value, unit):
    return value.lower() + unit
```

使用该过滤器：

```
{% load blogtrans %}

{{ 4|blogtrans:"$" }}
{{ '10'|blogtrans:'￥' }}
```

此时即使给过滤器传递数值类型也不会出现异常。

10.3.2 编写自定义标签

由于标签可以做更多的事情，因此开发自定义标签比开发过滤器更复杂。幸好 Django 为创建自定义标签提供了一些快捷方式，可以帮开发人员快速开发自定义标签。

其中，simple_tag() 是最简单的一类快捷方式，如下所示：

```
django.template.Library.simple_tag()
```

很多标签的工作就是接收一些参数并简单运算，最后返回运算结果，对于这种类型的标签可以使用 simple_tag 进行开发。

下面以格式化输出当前日期为例，查看如何使用 simple_tag：

```
import datetime
from django import template

register = template.Library()

@register.simple_tag
def current_time(format_string):
    return datetime.datetime.now().strftime(format_string)
```

在模板中应用：

```
{% load current_time %}
{% current_time "%y-%m-%d" %}
```

在 simple_tag 自定义标签中还可以使用模板的上下文对象，仍然以 about 视图的"today""yesterday"和"tomorrow"上下文对象为例，修改 current_time 标签方法：

```
import datetime
from django import template

register = template.Library()

@register.simple_tag(takes_context=True)
```

```
def current_time(context, format_string):
    return context['now'].strftime(format_string)
```

> 此时自定义标签方法的第一个参数必须是"context"。

修改模板文件 about.html：

```
{% load current_time %}

{% current_time "%Y-%m-%d" %}
```

输出结果如下图所示。

另外 simple_tag 还可以接收任意数量的位置参数和关键字参数：

```
@register.simple_tag
def my_tag(a, b, *args, **kwargs):
    warning = kwargs['warning']
    profile = kwargs['profile']
    ...
    return ...
```

10.4 模板继承

前面讲解了 Django 模板的基本结构与语法，本节进一步讲解模板的另一个重要使用方法——模板继承。

Django 的模板就像编程语言中的类一样是可以继承的，通过合理使用模板继承可以减少开发工作量，提高代码可复用性，进而提升工作效率。如果仔细观察绝大多数网站，就会发现无论网站规模有多大，一个网站的不同页面之间都会有相同的部分。以 W3School 为例，可以看到，该网站的头部、主菜单、底部网站信息在任何页面都是一样的，甚至可以认为左侧菜单、右侧广告栏以及中间主窗体都是一样的，只是填充的内容不同而已。

面对这样的网站，如果开发人员对每一个页面都单独开发，那么工作量会非常大，而一旦需要进行页面重构就会非常困难，需要修改所有页面。面对此类问题的通用做法就是提出页面的通用部分，进行单独开发，然后不同的页面继承这些公共部分，这样会在很大程度上减少维护成本。

Django 通过把网页中每一个通用部分定义在 block 中的方式实现了代码分离，下面以 Blog 应用程序为例演示模板继承。

首先修改 base.html 模板：

```
{% load static %}
<!DOCTYPE HTML>
<html>
<head>
    <meta http-equiv="Content-Type" content="text/html; charset=gb2312;" />
    <meta http-equiv="Content-Language" content="zh-cn" />
    <title>{% block title %}My amazing site{% endblock %}</title>
    <link rel="stylesheet" type="text/css" href="{% static 'css/myblog.css' %}" />
</head>

<body>
    <div id="nav">
        {% block menu %}
        <ul>
            <li><a href="/blog/">Home</a></li>
            <li><a href="/blog/about/">About</a></li>
        </ul>
        {% endblock %}
    </div>

    <div id="content">
        {% block content %}{% endblock %}
    </div>
</body>
</html>
```

它是网站的模板骨架，在这个模板骨架中没有定义任何代码实现，只定义了页面结构，该页面包括一个菜单栏（nav）以及页面主体（content）。

为了使网页结构看起来更直观，修改 myblog.css：

```
* {
    margin: 0;
    border: 0;
```

```css
        padding: 0;
        font-size: 13pt;
}

body {
    width:100%;
    height:100%;
    vertical-align:middle;
    text-align: center;
}

#nav {
    height: 40px;
    margin-bottom: 10px;
    border-top: #060 2px solid;
    border-bottom: #060 2px solid;
    background-color: #690;
}

#nav ul {
    list-style: none;
    margin-left: 50px;
}

#nav li {
    display: inline;
    line-height: 40px;
    float:left;
}

#nav a {
    color: #fff;
    text-decoration: none;
    padding: 20px 20px;
}

#nav a:hover {
    background-color: #060;
}

#content {
    margin-top: 10px;
    width: 1000px;
    height: 100%;
    display: inline-block;
    text-align: center;
}

.blog_head {
    background-color: rgb(45, 243, 243);
    border: 1px black solid;
    margin-bottom: 3px;
    text-align: left;
```

```
}
.blog_body {
    color: rgb(13, 27, 230);
    background-color: rgb(157, 158, 158);
    border: 1px black solid;
    min-height: 50px;
    margin-bottom: 10px;
    text-align: left;
}
```

此时重启 Web 服务查看网页显示效果，效果如下图所示。

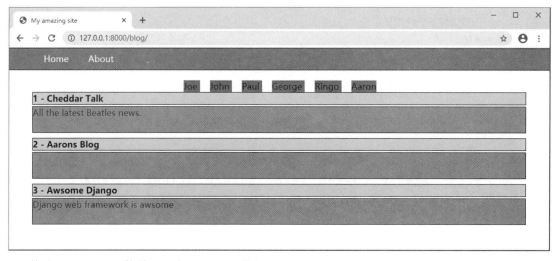

修改 about.html，使其继承自 base.html 模板：

```
{% extends "base.html" %}
{% load current_time %}

{% block content %}
```

<p> 类视图是 Django 框架提供的一系列预置视图，开发人员可以直接拿来使用，这极大地提高了代码开发效率。</p>

```
{% current_time "%Y-%m-%d" %}
{% endblock %}
```

刷新浏览器，单击"about"菜单查看网页显示效果，如下图所示。

可以看到网页保留了菜单栏以及网页主体。单击"Home"菜单又能返回首页。

注意

extends 标签必须位于文件的第一行，即使它前面是注释也不可以。

最终生成的 HTML 文件如下图所示。

```
<!DOCTYPE HTML>
<html>
<head>
    <meta http-equiv="Content-Type" content="text/html; charset=gb2312;" />
    <meta http-equiv="Content-Language" content="zh-cn" />
    <title>My amazing site</title>
    <link rel="stylesheet" type="text/css" href="/static/css/myblog.css" />
</head>
<body>
    <div id="nav">

        <ul>
            <li><a href="/blog/">Home</a></li>
            <li><a href="/blog/about/">about</a></li>
        </ul>

    </div>

    <div id="content">

<p>类视图是Django框架提供的一系列预置视图，开发人员可以直接拿来使用，这极大地提高了代码开发效率。</p>
2020-04-25


    </div>
</body>
</html>
```

在这里细心的读者可能会发现，about 模板中并没有定义 {% block menu %}…{% endblock %}，但是在最后生成的 HTML 文件里面却包含了这一部分内容，这就是模板继承的优势，基于此可以对项目进行细分，每个人专注自己的业务实现。

最后再做一个比较常见的网页跳转实验。前面在首页上列举了全部作者，在这里可以允许用户单击作者姓名跳转到作者详情页，修改 blog_list.html 模板，为 {{ author.name }} 增加动态 URL：

```
{% extends "base.html" %}
{% load static %}

{% block content %}

    {% for author in author_list %}
        <span style="background-color:grey; margin:10px;">
            <a href="{% url 'blog:author-detail' author.id %}">{{ author.name }}</a>
        </span>
    {% endfor %}

    {% for entry in entries %}
        <h2 class="blog_head">{{ entry.blog.id }} - {{ entry.blog.name }}</h2>
        <p class="blog_body">
            {{ entry.blog.tagline }}
        </p>
```

```
        {% endfor %}
{% endblock %}
```

此时单击作者姓名就能跳转到作者详情页，效果如下图所示。

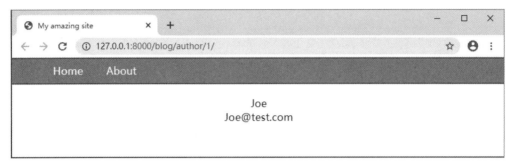

如果想在子模板中引用父模板中的 block，那么可以使用 {{ block.super }} 标签。例如，在上面例子中，想在菜单中添加更多的菜单，可以按照下面代码实现：

```
{% block menu %}
    {{ block.super }}
    <ul>
        <li><a href="/search/">Search</a></li>
    </ul>
{% endblock %}
```

刷新浏览器可以看到多了一个菜单项，如下图所示。

需要注意的是，在一个模板文件中 block 的名字是唯一的，不能重复。

Django 中的模板虽然可以无限继承，但是一般建议 3 层就可以了：

❏ Base.html 层，这里定义网站的整体骨架。
❏ 公共模块 .html，这里定义具体的模块，例如页眉、页脚等独立内容。
❏ 具体网页 .html，这里是真正实现具体功能的模板，会重写具体模块并继承 base.html。

总之根据自己的需求进行代码实现。

Django 官网给出了一些最佳实践，在开发网站的时候可以参考，如：

❏ 在 base.html 中尽可能多地使用 block。
❏ 如果在多个模板中都重复写了一些代码，那么就可以考虑将这些代码放在上级模板的 block 中。

第 11 章 表单系统

HTML 表单是 Web 应用的重要组成部分,可以用来接收用户输入信息,如提交注册信息、编辑个人资料等。

本书主要介绍 Django 的表单实现,至于 HTML 表单的作用与用法请参阅其他资料。

Django 的表单系统可以完成绝大多数表单工作,包括显示表单内容、编辑表单、表单验证、表单提交等,对于大多数开发人员来说使用 Django 表单会比自己开发更安全。

11.1 Form 类

Django 表单系统的核心是 Form 类。Form 类用于描述表单并决定表单如何工作。Form 类的字段对应 HTML 的 <input> 元素,每一个表单字段都是一个类,这些字段用于管理表单数据并在表单提交时进行数据验证。

接下来以 author_form 为例演示如何使用 Django 表单。前面介绍过使用 author_form.html 模板展示用户注册的功能,下面使用表单改写前面的代码。

首先编写一个新的表单文件 author_forms.py,文件内容如下:

```
from django import forms

class AuthorForm(forms.Form):
    name = forms.CharField(label='作者姓名', max_length=100)
    email = forms.EmailField(label='电子邮箱')
```

打开 author_view.py,删除视图类 AuthorCreate,新增视图方法 AuthorCreate():

```
from django.http import HttpResponseRedirect
from django.shortcuts import render
from django.urls import reverse_lazy
from .models import Author

from .author_forms import AuthorForm


def AuthorCreate(request):
    if request.method == 'POST':
        form = AuthorForm(request.POST)
        if form.is_valid():
            #保存作者
```

```
                author = Author(name=form.cleaned_data["name"],
            email=form.cleaned_data["email"], created_by=request.user.id)
                author.save()
                return HttpResponseRedirect(reverse_lazy('blog:success', args=[request.
POST['name']])) 
        else:
            form = AuthorForm()

        return render(request, 'blog/author/author_form.html', {'form': form})

    # 如果是 GET 请求,显示空表单,用于处理第一次访问表单
    else:
        form = NameForm()

    return render(request, 'name.html', {'form': form})
```

代码分析:

前面视图使用 cleaned_data 读取表单字段的值,这是因为不论表单提交的字段是什么类型的数据,一旦表单验证通过(调用 is_valid() 方法并返回 True),那么验证通过的表单数据将会被保存在 form.cleaned_data 字典中。虽然此时仍然可以使用 request.POST 或者 form.data 方式读取字段值,但是 Django 推荐使用 cleaned_data 字典。

重启浏览器,访问新增作者页面,显示效果如下图所示。

添加作者信息,然后提交,效果如下图所示。

到此为止,一个最简单的表单就完成了。

11.2 表单字段类型

11.1 节中使用 AuthorForm 重新实现了添加作者的页面,AuthorForm 包含两个字段,这两个字段分别使用了 CharField 和 EmailField 类型,对应的 HTML 元素是 <input type="text" name="name" maxlength="100" required=" " id="id_name"> 和 <input type="email" name="email" required=" " id="id_

email">。每个 Django 表单字段都对应一个 Widget 类，这个 Widget 类负责返回以上 HTML 字符串。

除了 CharField 和 EmailField 之外，Form 类还提供了几十种字段类型，每种字段类型分别返回不同的 HTML 元素，下面对这些类型进行简单介绍。如果需要更详细的表单字段介绍，可以参考 Django 官网：https://docs.djangoproject.com/zh-hans/3.0/ref/forms/fields/#built-in-field-classes。

1. BooleanField

Widget：CheckboxInput(<input type="checkbox" ...>)。

空值：False。

标准值：True、False。

验证：如果设置了 required=True，则验证字段值是否为 True（是否是选中状态）。

验证点：required。

2. CharField

Widget：TextInput(<input type="text" ...>)。

空值：empty_value。

标准值：字符串。

验证：如果设置了 max_length、min_length，则验证字段长度是否符合要求，否则不验证；如果 strip 属性为 True，那个字段值前后的空白符将会被删除。

验证点：required、max_length、min_length。

3. ChoiceField

Widget：Select(<select><option ...>...</select>)。

空值：""。

标准值：字符串。

验证：验证字段值是否存在。

验证点：required、invalid_choice。

4. DateField

Widget：DateInput(<input type="text" ...>)。

空值：None。

标准值：Python datetime.date 对象。

验证：验证字段值是否是正确的时间格式字符串、datetime.date 对象、datetime.datetime 对象。

验证点：required、invalid。

5. DateTimeField

Widget：DateInput(<input type="text" ...>)。

空值：None。

标准值：Python datetime.datetime 对象。

验证：验证字段值是否是正确的时间格式字符串、datetime.date 对象、datetime.datetime 对象。

验证点：required、invalid。

6. DecimalField

Widget：当 Field.localize=False 时，对应 NumberInput(<input type="number" ...>)，否则对应

TextInput(<input type="text" ...>)。

空值：None。

标准值：Python decimal 对象。

验证：验证字段值是否是数值类型，字段值前后的空白符会被忽略掉。

验证点：required、invalid、max_value、min_value、max_digits、max_decimal_places、max_whole_digits。

7. FileField

Widget：ClearableFileInput(<input type="file" ...>)。

空值：None。

标准值：包含文件内容与文件名的 UploadedFile 对象。

验证：空文件或者没有选择文件。

验证点：required、invalid、missing、empty、max_length。

8. FilePathField

Widget：Select(<select><option ...>...</select>)。

空值：None。

标准值：文件路径。

验证：选中的选项是否存在于下拉列表中。

验证点：required、invalid_choice。

重要参数说明如下：

参数 Path 接收一个绝对路径，下拉列表将显示路径下的文件或者文件夹；参数 recursive 默认值为 False，当设置为 True 时列表将同时显示全部子文件夹的内容；参数 match 指定一个正则表达式，Django 根据正则表达式查找并显示文件；参数 allow_files 用于指定是否显示路径下的文件，默认为 True；参数 allow_folders 用于指定是否显示路径下的文件夹，默认为 False；参数 allow_files 和 allow_folders 之间至少有一个必须为 True。

9. ImageField

Widget：ClearableFileInput(<input type="file" ...>)。

空值：None。

标准值：包含文件内容与文件名的 UploadedFile 对象。

验证：验证字段是否是空文件或者没有选择文件，如果包含文件，验证文件后缀是否被 Pillow 支持（使用 ImageField 字段类型之前必须保证系统安装了 Pillow）。

验证点：required、invalid、missing、empty、invalid_image。

10. IntegerField

Widget：当 Field.localize=False 时，对应 NumberInput(<input type="number" ...>)，否则对应 TextInput(<input type="text" ...>)。

空值：None。

标准值：Python integer 对象。

验证：验证字段值是否是一个整数。

验证点：required、invalid、max_value、min_value。

11. MultipleChoiceField

Widget：SelectMultiple(<select multiple="multiple">...</select>)。

空值：[](空列表)。

标准值：一组字符串。

验证：所有选中值存在于下拉列表中。

验证点：required、invalid_choice、invalid_list。

代码示例：

```
from django import forms

food = (('somestuff', 'beef'),
        ('otherstuff', 'eggs'),
        ('banana', 'bar'))

class AuthorForm(forms.Form):
    columns = forms.MultipleChoiceField(
        required=False,
        choices=food,
        widget=forms.CheckboxSelectMultiple)
```

输入代码，效果如下图所示。

12. EmailField

Widget：EmailInput(<input type= "email" ...>)。

空值：""。

标准值：字符串。

验证：使用正则表达式验证字段值是否是标准的邮件地址。

验证点：required、invalid。

13. FloatField

Widget：当 Field.localize=False 时对应 NumberInput(<input type="number" ...>)，否则对应 TextInput(<input type="text" ...>)。

空值：None。

标准值：Python 浮点数。

验证：验证字段值是否是浮点数。

验证点：required、invalid、max_value、min_value。

11.3 表单字段通用属性

1. required
默认情况下，所有的表单字段都是必填字段，因此，如果提交表单时没有为字段赋值则会抛出 ValidationError 异常。

对于非必填字段，可以设置 required=False 避免验证错误，例如：

```
forms.CharField(required=False)
```

2. label
为表单字段指定一个 label 元素用于显示字段信息，如前面的代码中 name 字段将会额外显示一个 label：

```
<label for="id_name">作者姓名:</label>
```

3. initial
为字段设置初始值。

4. help_text
为字段添加帮助性文字。

5. error_messages
重写字段的默认错误提示信息，error_messages 是一个字典类型。

例如，设置当 CharField 的 "required" 验证失败时显示 "请输入作者姓名"：

```
name = forms.CharField(error_messages={'required': '请输入作者姓名'})
```

6. localize
设置表单字段是否启用本地化。

7. disabled
当设置 disabled=True 时，使用 HTML disabled 属性禁用字段。

11.4 表单与模板

在模板中渲染表单非常简单，只需要在模板中添加 form 标签即可：{{ form }}。另外，可以使用以下方式对 form 格式进行设置：

- {{ form.as_table }} 使用 <tr> 标签显示表单字段，需要注意的是，表单不会自动生成 <table> 标签，必须手工在模板中添加 <table> 标签。
- {{ form.as_p }} 使用 <p> 标签显示表单字段。
- {{ form.as_ul }} 使用 <li> 标签显示表单字段，注意表单不会生成 <ul> 标签，必须手工在模板中添加 <ul> 标签。

除了使用 {{ form }} 自动生成表单外，还可以手工创建表单内容：

```
{# Include the hidden fields #}
{% for hidden in form.hidden_fields %}
{{ hidden }}
```

```
{% endfor %}
{# Include the visible fields #}
{% for field in form.visible_fields %}
    <div class="fieldWrapper">
        {{ field.errors }}
        {{ field.label_tag }} {{ field }}
    </div>
{% endfor %}
```

其中,"hidden_fields"是表单的隐藏字段,visible_fields是表单中可显示字段。

如果不需要判断表单字段是否显示,则可以直接遍历 form 对象:

```
{% for field in form %}
    <div class="fieldWrapper">
        {{ field.errors }}
        {{ field.label_tag }} {{ field }}
        {% if field.help_text %}
        <p class="help">{{ field.help_text|safe }}</p>
        {% endif %}
    </div>
{% endfor %}
```

第 12 章

中间件

中间件是 Django 请求/响应处理的钩子框架。它是一个轻量级的、低级的"插件"系统，用于全局改变 Django 的输入或输出。

每个中间件组件负责实现一些特定的功能。例如，Django 包含一个中间件组件 AuthenticationMiddleware，它使用会话将用户与请求关联起来。

下面简单介绍一下 Django 自带的部分中间件。

12.1 缓存中间件

缓存中间件英文名是 Cache middleware，用于缓存整个网站。Django 中的缓存中间件有两个：

- class UpdateCacheMiddleware
- class FetchFromCacheMiddleware

如果激活了这两个中间件，那么 Django 将会按照 CACHE_MIDDLEWARE_SECONDS 的设置对所有基于 Django 开发的页面进行缓存。其中，UpdateCacheMiddleware 是响应阶段的中间件，用于更新缓存；FetchFromCacheMiddleware 是请求阶段的中间件，用于提取已缓存的页面。

需要注意的是，使用以上缓存中间件时，UpdateCacheMiddleware 和 FetchFromCacheMiddleware 必须成对出现，而且 UpdateCacheMiddleware 必须放在列表的起始第一个位置，FetchFromCacheMiddleware 放在列表的最后一个位置。settings.py 的设置如下：

```
MIDDLEWARE = [
    'django.middleware.cache.UpdateCacheMiddleware',
    '其他中间件',
    'django.middleware.cache.FetchFromCacheMiddleware'
]
```

下面是使用缓存中间件的一些注意事项：

- 只有状态码为 200 的 GET 或 HEAD 请求被高速缓存。
- 每个页面被缓存的秒数由响应的缓存控制（Cache-Control）的 max-age 决定，如果没有设置 max-age，那么页面的缓存时间由 CACHE_MIDDLEWARE_SECONDS 决定。
- HEAD 请求与 GET 请求接收相同的响应。
- 如果发生故障，将会从 process_request 返回一个原始响应的浅复制对象。
- 页面基于响应的 Vary 头进行缓存。
- 这个中间件还会设置响应的 Etag、Last-Modified、Expires 和 Cache-Control 头。

12.2 通用中间件

通用中间件 class CommonMiddleware 为完美主义者提供了一些便利的工具，例如：
- 禁止用户访问配置项 DISALLOWED_USER_AGENTS 中设置的代理服务，DISALLOWED_USER_AGENTS 可以是一组已经编译好的 Python 正则表达式对象。
- 基于配置文件中的 APPEND_SLASH 和 PREPEND_WWW 重写 URL。
- 为 non-streaming 响应设置 Content-Length。

12.3 GZip 中间件

对应的 Python 类：class GZipMiddleware。

> **注意**
> 由于研究人员发现压缩网页可能会导致网站受到攻击，因此使用 GZip 中间件时应十分谨慎。

GZip 中间件为支持 GZip 压缩的浏览器压缩网页内容（事实上所有主流浏览器都支持 GZip 压缩）。

GZip 中间件放在所有需要读写响应体的中间件之前，以便能够正确压缩网页内容。

需要注意的是，当以下任意一项为 True 时，GZip 中间件都不会对网页进行压缩：
- 网页内容小于 200 个字节。
- 响应已经设置了 Content-Encoding 头。
- 浏览器请求未发送包含 gzip 的 Accept-Encoding 头。

如果响应有一个 ETag 头，则 ETag 被弱化以符合 RFC 7232#section-2.1，Etag 是 URL 的 Entity Tag，用于标识 URL 对象是否改变，作用和 Last-Modified 相似。

使用 gzip_page() 装饰器可以为单个视图设置 GZip 压缩。

12.4 有条件的 GET 中间件

对应的 Python 类：class ConditionalGetMiddleware。

该中间件用于处理有条件的 GET 操作。

如果响应没有 ETag 头，该中间件将会在必要的时候为响应添加一个 ETag 头。如果响应包含 ETag 或者 Last-Modified，同时请求包含 If-None-Match 或者 If-Modified-Since，那么响应会被 HttpResponseNotModified 替代。

12.5 语言环境的中间件

对应的 Python 类：class LocaleMiddleware。

使用 LocaleMiddleware，可以根据请求的数据方便地为不同用户设置语言及网页内容。

12.6 消息中间件

对应的 Python 类：class MessageMiddleware。
启用基于 Cookie 和 session 的消息。

12.7 安全中间件

对应的 Python 类：class SecurityMiddleware。
为请求/响应过程提供多个安全改善方案，如果条件允许，建议启用该中间件，这样即使一些请求不是来自 Django 应用（如静态文件或者用户上传的文件）也会得到保护。
具体设置可参考 5.9 节。

12.8 会话中间件

对应的 Python 类：class SessionMiddleware。
启用对会话的支持。

12.9 站点中间件

对应的 Python 类：class CurrentSiteMiddleware。
向所有请求的 HttpRequest 对象中添加一个 site 属性，该属性代表当前站点。

12.10 身份验证中间件

Python 类：class AuthenticationMiddleware。
身份验证中间件会为每个请求的 HttpRequest 对象添加一个 user 属性，该属性代表当前登录的用户。

Python 类：class RemoteUserMiddleware。
该中间件使得 Django 能够使用 Web 服务器提供的身份认证功能，如 Web 服务器设置了 REMOTE_USER 环境变量。这类身份认证通常应用于企业内部网络，如 IIS 的集成 Windows 身份认证等。

当 Web 服务器处理身份认证时，通常会设置一个 REMOTE_USER 环境变量，以便在应用程序中使用。在 Django 的 request.META 属性中可以使用 REMOTE_USER。

使用 REMOTE_USER 需要先配置 MIDDLEWARE：

```
MIDDLEWARE = [
    '...',
    'django.contrib.auth.middleware.AuthenticationMiddleware',
    'django.contrib.auth.middleware.RemoteUserMiddleware',
    '...',
]
```

然后配置 AUTHENTICATION_BACKENDS：

```
AUTHENTICATION_BACKENDS = [
    'django.contrib.auth.backends.RemoteUserBackend',
]
```

此时 RemoteUserMiddleware 中间件就可以检索到 request.META['REMOTE_USER'] 中的用户名，然后自动验证并使用该用户登录网站。

注意，使用 RemoteUserMiddleware 时，如果在 Web 服务器中没有设置 REMOTE_USER，就不能登录该用户。另外，Django 管理后台自带的用户管理功能以及 createsuperuser 命令都不能处理 REMOTE_USER。

Python 类：class PersistentRemoteUserMiddleware。

仅在登录页面使用 Web 服务器提供的身份认证。

前面的 RemoteUserMiddleware 中间件从 HTTP 请求中获取 REMOTE_USER，该用户身份将会被用于所有页面请求。某些情况下，前端服务器只处理部分登录 URL 的身份验证工作，后续将由应用程序自己维护会话信息。PersistentRemoteUserMiddleware 解决了以上问题，它将持续记录会话信息直至用户退出。

12.11 CSRF 保护中间件

对应的 Python 类：class CsrfViewMiddleware。

通过添加一个隐藏的表单字段的方式提供了对跨站点请求伪造攻击的保护。

12.12 X-Frame-Options 中间件

对应的 Python 类：class XFrameOptionsMiddleware。

通过使用 X-Frame-Options 头提供了对点击劫持的保护。

12.13 中间件排序

由于很多 Django 中间件的执行都需要前置条件，因此应特别注意每个中间件的位置，下面对不同 Django 中间件类的排序进行说明。

- SecurityMiddleware：如果要启用 SSL，通常将该中间件放置在中间件列表的最上面，因为这样可以避免执行很多不必要的中间件。
- UpdateCacheMiddleware：放在任何能够修改 Vary 头的中间件前面，如 Session Middleware、GzipMiddleware、LocaleMiddleware。
- GzipMiddleware：放在任何可能修改或使用响应请求体的中间件前面。
  放在 UpdateCacheMiddleware 后面。
- SessionMiddleware：放在任何可能抛出异常并调用异常视图的中间件前面，例如 PermissionDenied 异常。
  放在 UpdateCacheMiddleware 后面。
- ConditionalGetMiddleware：放在任何可能修改响应的中间件前面（因为 ConditionalGetMiddleware 会设置 ETag 头）。
  放在 GZipMiddleware 后面，这样在压缩报文时就不会计算 ETag 头。

- LocaleMiddleware：最高优先级的一个中间件，放在 SessionMiddleware 和 UpdateCache Middleware 后面。
- CommonMiddleware：由于 CommonMiddleware 会设置 Content-Length 头，因此要将 CommonMiddleware 放在任何可能会修改响应的中间件前。如果任何中间件放置在 CommonMiddleware 前面同时修改了响应，那么必须重新计算 Content-Length。

  尽可能将 CommonMiddleware 放在顶部，因为当 APPEND_SLASH 或者 PREPEND_WWW 为 True 时可能会发生重定向。

  如果使用了 CSRF_USE_SESSIONS，那么还需要放置在 SessionMiddleware 的后面。
- CsrfViewMiddleware：放在任何需要处理 CSRF 攻击的中间件前。

  放在 RemoteUserMiddleware 或者任何执行登录操作的身份认证中间件前。

  如果使用了 CSRF_USE_SESSIONS，那么还需要放置在 SessionMiddleware 的后面。
- AuthenticationMiddleware：由于需要使用 sesstion，因此将 AuthenticationMiddleware 放在 SessionMiddleware 后面。
- MessageMiddleware：为了能够使用基于 session 的数据，需要将 MessageMiddleware 放在 SessionMiddleware 之后。
- FetchFromCacheMiddleware：放在任何可能修改 Vary 头的中间件后面。
- FlatpageFallbackMiddleware：尽可能地放在最后。
- RedirectFallbackMiddleware：尽可能地放在最后。

12.14 开发中间件

中间件是一个钩子，可以加入 Django 的请求或响应过程中并执行一些特殊方法。

1. 方法中间件

示例代码如下：

```
def simple_middleware(get_response):
    # One-time configuration and initialization.

    def middleware(request):
        # Code to be executed for each request before
        # the view (and later middleware) are called.

        response = get_response(request)

        # Code to be executed for each request/response after
        # the view is called.

        return response

    return middleware
```

2. 类中间件

示例代码如下：

```
class SimpleMiddleware:
```

```python
def __init__(self, get_response):
    self.get_response = get_response
    # One-time configuration and initialization.

def __call__(self, request):
    # Code to be executed for each request before
    # the view (and later middleware) are called.

    response = self.get_response(request)

    # Code to be executed for each request/response after
    # the view is called.

    return response
```

其中，get_response 是 Django 提供的一个可调用对象（callable），当中间件是列表中的最后一项时，get_response 是真正的视图，否则 get_response 是下一个中间件。当前中间件并不需要知道 get_response 究竟是什么。

中间件文件可以放在任何 Python 可以访问的位置。

3. 激活中间件

要想激活一个中间件，只需要将中间件的 Python 路径加入 MIDDLEWARE 列表中即可。

第 13 章
自动化测试

任何软件产品在发布之前都应该进行测试，充分合理的测试工作可以保证产品质量。现代软件产品的规模越来越庞大、业务逻辑越来越复杂、发布周期越来越短，如果仅仅使用人工测试，这会带来巨大的工作量以及人员成本，自动化测试可以在一定程度上改善以上问题。

13.1 编写第一个测试用例

为了演示 Django 是如何进行自动化测试的，修改 polls\models.py，为 Question 模型添加一个判断问卷发布日期的方法 was_published_recently()，如果问卷是最近一天内发布的则返回 True。

```python
import datetime
from django.utils import timezone
...

class Question(models.Model):
    question_text = models.CharField(max_length=200)
    pub_date = models.DateTimeField('date published')
    description = models.CharField(max_length=200, null=True, blank=True)
...

    def was_published_recently(self):
        return self.pub_date >= timezone.now() - datetime.timedelta(days=1)
```

仔细观察 was_published_recently() 方法，可以发现该方法把所有发布日期晚于昨天（使用当前时间减去 datetime.timedelta(days=1) 表示一天前）的问卷都认为是最近发布的问卷，这里就会存在一个问题，如果某一个问卷的发布日期定在将来时间，也就是说用户希望推迟发布问卷，该方法也会返回 True，这是不准确的。下面根据该场景进行测试。

Django 的自动化测试代码通常会放在一个以 test 开头的 Python 脚本文件中，Django 系统会根据文件名来查找测试代码。在使用 Django 命令行创建应用程序的时候，系统已经创建了一个叫作 tests.py 的自动化测试脚本文件。

将以下测试代码复制到 polls\ tests.py 脚本中：

```python
#!/usr/bin/python
# -*- coding: UTF-8 -*-

import datetime
```

```python
from django.utils import timezone
from django.test import TestCase

from .models import Question

class QuestionModelTests(TestCase):

    def test_was_published_recently_with_future_question(self):
        """
        当问卷的发布日期是未来的某一天时,was_published_recently()方法将会返回False。
        """
        time = timezone.now() + datetime.timedelta(days=30)
        future_question = Question(pub_date=time)
        self.assertIs(future_question.was_published_recently(), False)
```

13.2 执行测试用例

执行应用程序下的全部测试用例:

```
$ python manage.py test polls
```

执行应用程序下指定模块的测试用例:

```
$ python manage.py test polls.tests
```

执行应用程序下指定测试类的测试用例:

```
$ python manage.py test polls.tests.QuestionModelTests
```

执行具体某一个测试用例:

```
$ python manage.py test polls.tests.QuestionModelTests.test_was_published_recently_with_future_question
```

执行某一文件夹下的全部测试用例:

```
$ python manage.py test unittest/
```

模糊匹配测试用例:

```
$ python manage.py test --pattern="tests_*.py"
$ python manage.py test -p ="tests_*.py"
```

命令分析:
- python manage.py test 检索应用程序中的测试用例。
- 如果发现自动化测试类(django.test.TestCase 的子类),则创建一个测试数据库。
- 检索测试方法(测试方法名以 test 开头)。
- 执行测试方法并输出测试结果。
- 删除测试数据库。

13.3 修改代码中的 bug

找到代码中存在的 bug 后就需要更正它。在前面代码中,如果问卷的发布日期是将来的某一天,was_published_recently() 方法会返回 False,对其修改如下:

```
def was_published_recently(self):
    now = timezone.now()
    return now - datetime.timedelta(days=1) <= self.pub_date <= now
```

重新执行测试用例,效果如下图所示。

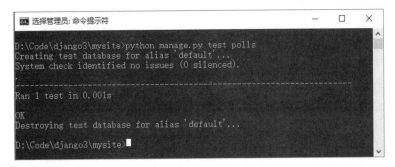

13.4 边界值测试

边界值测试是用来测试用户输入边界值时代码功能是否正常的一种测试方式,边界值也是最经常出现异常的情况。

前面示例将当前时间与前一天之间所发布的问卷定义为"最近发布的问卷",那么它的边界值是"比前一天早 1 秒"和"比前一天晚 1 秒",据此添加两个新的测试用例:

```
def test_was_published_recently_with_old_question(self):
    """
    当问卷的发布日期比前一天还早 1 秒时,was_published_recently() 返回 False。
    """
    time = timezone.now() - datetime.timedelta(days=1, seconds=1)
    old_question = Question(pub_date=time)
    self.assertIs(old_question.was_published_recently(), False)

def test_was_published_recently_with_recent_question(self):
    """
    当问卷的发布日期比前一天晚 1 秒时,was_published_recently() 返回 True。
    """
    time = timezone.now() - datetime.timedelta(hours=23, minutes=59, seconds=59)
    recent_question = Question(pub_date=time)
    self.assertIs(recent_question.was_published_recently(), True)
```

执行测试用例,效果如下图所示。

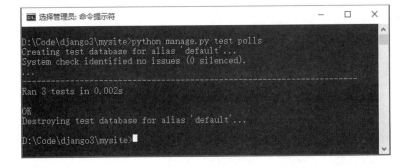

从输出结果可以看到，本次一共执行了 3 个测试用例，全部测试通过（图中的 OK 表示全部通过）。

13.5 测试自定义视图

目前 polls 应用程序可以使用任何 pub_date 来发布问卷，甚至使用未来的某一天。为了解决这一问题，将问卷的显示逻辑修改为：如果问卷的发布日期还没到，则问卷不可见。修改 IndexView 视图的 get_queryset() 方法：

```python
from django.utils import timezone
from django.views import generic
...

class IndexView(generic.ListView):
    template_name = 'polls/index.html'
    context_object_name = 'latest_question_list'

    def get_queryset(self):
        """返回最近发布的 5 个调查问卷。"""
        return Question.objects.filter(
            pub_date__lte=timezone.now()
        ).order_by('-pub_date')[:5]
```

修改 URL：

```python
path('', views.IndexView.as_view(), name='index'),
```

为了对新视图进行测试，需要分别创建：发布日期早于当前时间的问卷、发布日期晚于当前时间的问卷和多种发表时间混合的问卷。

修改 polls\tests.py 脚本，添加 create_question() 用于创建测试问卷：

```python
def create_question(question_text, days):
    """
    使用指定的文本和日期数量创建问卷。
    如果日期数量是负数，则表示发布日期早于当前时间，
    如果日期数量是正数，则表示发布日期晚于当前时间。
    """
    time = timezone.now() + datetime.timedelta(days=days)
    return Question.objects.create(question_text=question_text, pub_date=time)
```

继续在 polls\tests.py 中添加视图测试代码：

```python
from django.urls import reverse
...
class QuestionIndexViewTests(TestCase):
    def test_no_questions(self):
        """
        如果当前系统没有符合条件的问卷则返回提示信息。
        """
        response = self.client.get(reverse('polls:index'))
        self.assertEqual(response.status_code, 200)
        self.assertContains(response, u"还没有调查问卷！")
        self.assertQuerysetEqual(response.context['latest_question_list'], [])
```

```python
def test_past_question(self):
    """
    早于当前时间发表的问卷将会显示在 index 页面上。
    """
    create_question(question_text="Past question.", days=-30)
    response = self.client.get(reverse('polls:index'))
    self.assertQuerysetEqual(
        response.context['latest_question_list'],
        ['<Question: Past question.>']
    )

def test_future_question(self):
    """
    晚于当前时间发表的问卷将不会显示在 index 页面上。
    """
    create_question(question_text="Future question.", days=30)
    response = self.client.get(reverse('polls:index'))
    self.assertContains(response, u"还没有调查问卷！")
    self.assertQuerysetEqual(response.context['latest_question_list'], [])

def test_future_question_and_past_question(self):
    """
    系统中同时存在已经发表和还未发表的问卷，只显示已经发表的问卷。
    """
    create_question(question_text="Past question.", days=-30)
    create_question(question_text="Future question.", days=30)
    response = self.client.get(reverse('polls:index'))
    self.assertQuerysetEqual(
        response.context['latest_question_list'],
        ['<Question: Past question.>']
    )

def test_two_past_questions(self):
    """
    index 页面可以同时显示多个问卷。
    """
    create_question(question_text="Past question 1.", days=-30)
    create_question(question_text="Past question 2.", days=-5)
    response = self.client.get(reverse('polls:index'))
    self.assertQuerysetEqual(
        response.context['latest_question_list'],
        ['<Question: Past question 2.>', '<Question: Past question 1.>']
    )
```

代码分析：
- 测试用例使用 create_question() 创建测试数据。
- 测试用例使用 self.client.get 模拟用户通过浏览器访问调查问卷（self.client 继承自基类 TestCase）。
- 通过 response.context 对象获取网络请求返回的数据。

执行测试用例，效果如下图所示。

13.6 测试 DetailView

虽然 index 视图屏蔽了用户对未发布问卷的访问权限，但是，如果用户知道或者根据一定的规律猜到问卷的 ID，那么他可以通过这个 ID 访问问卷的详细信息（通过拼写 URL 的方式）。为了解决这个问题，按照下面方式修改 DetailView：

```python
class DetailView(generic.DetailView):
    model = Question
    template_name = 'polls/detail.html'

    def get_queryset(self):
        """
        只提取已经发表的问卷。
        """
        return Question.objects.filter(pub_date__lte=timezone.now())
```

修改 URL：

```python
path('<int:pk>/', views.DetailView.as_view(), name='detail'),
```

添加测试用例：

```python
class QuestionDetailViewTests(TestCase):
    def test_future_question(self):
        """
        如果被查询的问卷还没有发表则返回 404 错误。
        """
        future_question = create_question(question_text='Future question.', days=5)
        url = reverse('polls:detail', args=(future_question.id,))
        response = self.client.get(url)
        self.assertEqual(response.status_code, 404)

    def test_past_question(self):
        """
        如果被查询的问卷已经发表了则返回问卷内容。
        """
        past_question = create_question(question_text='Past Question.', days=-5)
        url = reverse('polls:detail', args=(past_question.id,))
```

```
        response = self.client.get(url)
        self.assertContains(response, past_question.question_text)
```

执行测试用例，效果如下图所示。

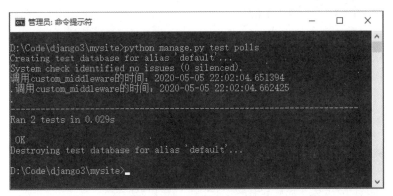

与 DetailView 相似，ResultsView 也可以按照上面方法添加 get_queryset() 和测试类。另外，对于还没有选项的问卷，可以选择不显示在 index 页面和 result 页面上。针对这些规则可以添加更多的测试用例。

对于不同权限的用户，index 页面所显示的信息可能不同，这些都可以编写自动化测试用例。

建议开发人员每完成一个新的功能都要编写相应的自动化测试用例，虽然当测试用例很多时代码会缺失美感，但是比较而言代码质量更重要。

下面是一些测试用例的最佳实践：

❑ 为每一个模型或视图单独创建测试类。
❑ 每一个测试用例只用来测试一种情况。
❑ 测试用例名要能够清晰地解释测试用例。

第 14 章

国际化和本地化

国际化与本地化的目的是为了适应多语言环境，同一个网站应该能够在不同的语言环境下显示不同的语言内容。

Django 完全支持翻译文本，格式化日期、时间、数字以及时区。

本质上，Django 的国际化做了两件事：

- 它允许开发者和模板作者指定应用程序的哪个部分应该被翻译或格式化为本地语言，显示效果应该符合本地文化。
- 它根据特定用户的偏好，使用钩子来本地化 Web 应用。

Django 的翻译工作是基于目标语言完成的，格式化是依赖目标国家完成的。而这些信息都是通过浏览器的 Accept-Language 头得到的。在这里时区并不是必然可用的。

14.1 名词解释

1. 国际化

国际化（internationalization）是设计一个适用于多种语言和地区的应用程序的过程。适用于多种语言和地区的含义是当使用不同语言及处于不同地区的用户在使用这个应用程序时，应用程序必须使用他们能看懂的语言和符合他们的文化习惯来显示信息。国际化有时候被简称为 I18N，因为 internationalization 的 i 和 n 之间有 18 个字母。

2. 本地化

本地化（localization）是指通过增加本地描述的构件（locale-specific component）和文字翻译工作来使应用程序适应于不同的语言和地区的过程。本地化有时候被简称为 L10N，因为 localization 的字母 l 和 n 之间有 10 个字母。通常本地化最耗时的工作是文字翻译。本地化工作者们要根据地区的具体需求来为日期、数字和货币等数据建立新的格式。其他类型的数据，像声音、图像等，也需要根据具体体需求来决定是否本地化。

3. 区域设置名称

区域设置名称（locale name）即可以是 ll 格式的语言规范，也可以是 ll_CC 格式的语言＋国家规范，如 it、de_AT、es、pt_BR。注意在区域设置名称中代表语言的部分永远使用小写英文字母，代表国家的部分永远使用大写英文字母。语言与国家之家使用下画线分隔。

4. 语言代码

语言代码（language code）表示语言名字。浏览器接收 HTTP 报文的 Accept-Language 头并发送对应的语言名，如 it、de-at、es、pt-br。语言代码通常使用小写字母，但是 Accept-Language 头不区分字母大小写。语言代码以中横线分隔。

5. 消息文件

消息文件（message file）是纯文本文件，表示具体某一种语言的翻译文本。消息文件包含所有可用的翻译字符串，以及它们在给定语言中应该如何表示。消息文件的扩展名为 .po。

6. 翻译字符串

翻译字符串（translation string）表示可以翻译的文字。

7. 格式化文件

格式文件（format file）是一个 Python 模块，它定义了给定地区的数据格式。

14.2 翻译概述

为了使 Django 项目能够被翻译，需要在 Python 代码以及 Django 模板中添加一些钩子。这些钩子叫作翻译字符串。这些钩子会告诉 Django：如果当前语言存在文本翻译，那么将文本翻译成用户能看懂的文字。开发人员需要标注出可翻译的字符串，Django 系统只翻译它能找到的可翻译字符串。

Django 提供了一些可以提取翻译字符串到消息文件（message file）的工具。消息文件为翻译人员提供了一种将翻译字符串翻译成目标语言的方便方式。一旦翻译人员填充好消息文件，这些消息文件就必须经过编译才能使用。这个过程由 GNU gettext 完成。

一旦完成以上所有步骤，Django 就会接管剩下的工作，它会根据访问用户的语言在线翻译网站内容。

默认情况下，Django 的国际化钩子是开启的，这会造成一定的性能开销，如果系统不需要进行国际化，可以在配置文件中关掉它：USE_I18N = False。

14.3 在 Python 中进行国际化

Python 标准库使用 gettext() 方法生成翻译字符串，为了书写方便，Python 会使用 "_()" 作为 gettext() 的别名注册到全局变量中。但是，在 Django 中不遵守这种命名方式，这是因为有时开发人员会使用 gettext_lazy() 作为翻译方法，另外，"_" 在 Python 的交互窗口中又代表了 "上一步操作结果"。为了避免冲突，Django 建议显式地以 "_" 为名导入 gettext()。

下面是 Python 对 gettext() 方法的官方说明：

```
gettext.gettext(message)
    Return the localized translation of message, based on the current global domain,
language, and locale directory. This function is usually aliased as _() in the local
namespace (see examples below).
```

下面是 gettext 的常见用法：

```
import gettext
gettext.bindtextdomain('myapplication', '/path/to/my/language/directory')
gettext.textdomain('myapplication')
```

```
_ = gettext.gettext
# ...
print(_('This is a translatable string.'))
```

下面以调查问卷首页为例演示如何在视图方法中添加翻译字符串。

首先在 polls\templates\polls\index.html 中增加标题：

```
<h1>{{title}}</h1>
```

修改 polls/views.py，使 IndexView 可以翻译 title：

```
from django.utils.translation import gettext as _

class IndexView(generic.ListView):
    template_name = 'polls/index.html'
    context_object_name = 'latest_question_list'

    def get_queryset(self):
        """返回最近发布的 5 个调查问卷。"""
        return Question.objects.filter(
            pub_date__lte=timezone.now()
        ).order_by('-pub_date')[:5]

    def get_context_data(self, **kwargs):
        ''' 翻译 '''
        context = super().get_context_data(**kwargs)
        context['title'] = _("Question List")
        return context
```

对于变化的文本可以使用占位符进行翻译，如日期文本：

```
def my_view(request, m, d):
    output = _('Today is %(month)s %(day)s.') % {'month': m, 'day': d}
    return HttpResponse(output)
```

英语环境将翻译为"Today is November 26."，而西班牙语环境将翻译为"Hoy es 26 de Noviembre."。基于这个原因，在翻译字符串中最好使用命名字符串，如 %(month)，而不是位置参数，如 %s。

需要注意的是，不要在翻译字符串中使用变量或者表达式，例如：

```
def my_view(request):
    sentence = 'Welcome to my site.'
    output = _(sentence)
    return HttpResponse(output)
```

或者：

```
def my_view(request):
    words = ['Welcome', 'to', 'my', 'site.']
    output = _(' '.join(words))
    return HttpResponse(output)
```

这是因为 django-admin makemessages 不能从变量或者表达式中查找翻译字符串。

14.3.1 注释

对于任何翻译字符串都应该给出合适的注释，注释以 Translators 开头，如：

```python
def get_context_data(self, **kwargs):
    context = super().get_context_data(**kwargs)
    # Translators: 调查问卷首页标题
    context['title'] = _("Question List")
    return context
```

执行以下命令生成消息文件：

```
django-admin makemessages -l zh_CN
```

注释文本将会显示在 .po 文件中，如下图所示。

```
40  #. Translators: 调查问卷首页标题
41  #: .\polls\views.py:70
42  msgid "Question List"
43  msgstr ""
```

同样模板中的翻译字符串注释也会显示在 .po 文件中：

```
{% comment %}Translators: View verb{% endcomment %}
{% trans "View" %}

{% comment %}Translators: Short intro blurb{% endcomment %}
<p>{% blocktrans %}A multiline translatable
literal.{% endblocktrans %}</p>
```

14.3.2 空操作

使用 django.utils.translation.gettext_noop() 方法可以标记一个字符串为翻译字符串，但是并不立即翻译它。

14.3.3 复数

django.utils.translation.ngettext() 方法可以对翻译字符串进行单复数处理，该方法接收 3 个参数：字符串的单数形式、字符串的复数形式和对象出现次数。这种方法在类似英语这种区分单复数形式的语言中非常有用。下面根据调查问卷数量进行区别输出：

```python
from django.utils.translation import ngettext

class IndexView(generic.ListView):
    ...
    def get_context_data(self, **kwargs):
        '''翻译'''
        ...
        count = len(context['latest_question_list'])
        context['count'] = ngettext(
            'there is %(count)d question',
            'there are %(count)d questions',
            count) % {
            'count': count,
        }
        return context
```

执行 makemessages 命令更新 .po 文件，如下图所示。

```
50 #: .\polls\views.py:77
51 #, python-format
52 msgid "there is %(count)d question"
53 msgid_plural "there are %(count)d questions"
54 msgstr[0] ""
55 msgstr[1] ""
```

由于在不同语言中,对于事物名字的单复数表示是不同的,这有些复杂,因此不建议开发人员自己处理,可以考虑使用 verbose_name,例如:

```python
text = ngettext(
    'There is %(count)d %(name)s object available.',
    'There are %(count)d %(name)s objects available.',
    count
) % {
    'count': count,
    'name': Report._meta.verbose_name,
}
```

14.3.4 上下文标记

很多时候一个单词会有多种解释,如英语中的"May",即可以表示"五月",也可以表示一个动词。为了让翻译人员能够根据上下文准确翻译这些单词,可以使用 django.utils.translation.pgettext() 方法,对于复数形式可以使用 django.utils.translation.npgettext() 方法,这两个方法接收的第一个参数都是上下文字符串。

在生成的 .po 文件中,每个翻译字符串有多少上下文标记就会生成多少个字符串(其中上下文用 msgctxt 表示)。

以单词"May"为例演示上下文标记的用法,首先修改视图代码:

```python
from django.utils.translation import pgettext

class IndexView(generic.ListView):
    ...
    def get_context_data(self, **kwargs):
        context = super().get_context_data(**kwargs)
        # Translators: 英文单词May--月份
        context['month'] = pgettext("month name", "May")
        # Translators: 英文单词May--动词
        context['verb'] = pgettext("verb", "May")
        return context
    ...
```

执行以下命令生成 .po 文件:

```
django-admin makemessages -l zh_CN
```

用记事本打开 .po 文件,如下图所示。

```
57 #. Translators: 英文单词May--月份
58 #: .\polls\views.py:84
59 msgctxt "month name"
60 msgid "May"
61 msgstr ""
62
63 #. Translators: 英文单词May--动词
64 #: .\polls\views.py:86
65 msgctxt "verb"
66 msgid "May"
67 msgstr ""
```

14.3.5 延迟翻译

延迟翻译(lazy translation)可以使翻译字符串在被使用的时候才会取得翻译结果。相关方法的命名规则为：普通翻译方法名 + "_lazy"后缀，如 gettext_lazy。

延迟翻译方法为字符串赋值一个延迟引用，只有当字符串被使用时才会取得最终翻译结果，如模板渲染时。

如果翻译方法在模块加载时被执行，如定义模型、表单、模型表单时，使用延迟翻译非常有必要。

下面列举几种使用延迟翻译的情形。

1. 模型字段的 verbose_name 和 help_text 属性

```python
from django.db import models
from django.utils.translation import gettext_lazy as _

class MyThing(models.Model):
    name = models.CharField(help_text=_('This is the help text'))
```

或者：

```python
class MyThing(models.Model):
    kind = models.ForeignKey(
        ThingKind,
        on_delete=models.CASCADE,
        related_name='kinds',
        verbose_name=_('kind'),
    )
```

2. 模型的 verbose_name

```python
from django.db import models
from django.utils.translation import gettext_lazy as _

class MyThing(models.Model):
    name = models.CharField(_('name'), help_text=_('This is the help text'))

    class Meta:
        verbose_name = _('my thing')
        verbose_name_plural = _('my things')
```

3. 模型方法的 short_description 属性

```python
from django.db import models
from django.utils.translation import gettext_lazy as _

class MyThing(models.Model):
    kind = models.ForeignKey(
        ThingKind,
        on_delete=models.CASCADE,
        related_name='kinds',
        verbose_name=_('kind'),
    )
```

```python
    def is_mouse(self):
        return self.kind.type == MOUSE_TYPE
    is_mouse.short_description = _('Is it a mouse?')
```

Django 代码中延迟翻译方法的返回值可以像普通字符串一样使用，但是不能在其他 Python 代码中使用。例如 request 对象中不能使用延迟翻译方法：

```python
body = gettext_lazy("I \u2764 Django")  # (unicode :heart:)
requests.post('https://example.com/send', data={'body': body})
```

为了避免出现这类问题，可以在使用延迟翻译方法时进行转换：

```python
requests.post('https://example.com/send', data={'body': str(body)})
```

对复数名词来说，使用延迟翻译也会存在翻译时不知道名字个数的情况，此时也可以使用数量参数进行判断：

```python
from django import forms
from django.utils.translation import ngettext_lazy

class MyForm(forms.Form):
    error_message = ngettext_lazy("You only provided %(num)d argument",
        "You only provided %(num)d arguments", 'num')

    def clean(self):
        # ...
        if error:
            raise forms.ValidationError(self.error_message % {'num': number})
```

当翻译字符串只有一个未命名的占位符时，可以简化代码，此时将会直接传入 number 参数：

```python
class MyForm(forms.Form):
    error_message = ngettext_lazy(
        "You provided %d argument",
        "You provided %d arguments",
    )

    def clean(self):
        # ...
        if error:
            raise forms.ValidationError(self.error_message % number)
```

格式化延迟翻译字符串时可以使用 django.utils.text.format_lazy() 方法：

```python
from django.utils.text import format_lazy
from django.utils.translation import gettext_lazy
...
name = gettext_lazy('John Lennon')
instrument = gettext_lazy('guitar')
result = format_lazy('{name}: {instrument}', name=name, instrument=instrument)
```

14.3.6 本地化的语言名

使用 get_language_info() 方法可以获取语言的详细信息。下面在 Django 交互窗口查看语音 de 的详细信息，首先执行 shell 命令进入 Django 控制台：

```
django-admin shell --settings=mysite.settings --pythonpath D:\Code\django3\mysite
```

接下来调用 get_language_info() 方法查看语言信息：

```
>>> from django.utils.translation import activate, get_language_info
>>> activate('fr')
>>> li = get_language_info('de')
>>> print(li['name'], li['name_local'], li['name_translated'], li['bidi'])
German Deutsch Allemand False
```

字典 li 的 name、name_local 和 name_translated 属性分别表示语言 de 在英语、de 自身以及当前活动语言中的名字。bidi 表示是否为双向语言。

前面使用 activate() 激活 fr 作为当前语言，如果换成中文，li 的输出结果就不一样了：

```
>>> activate('zh')
>>> li = get_language_info('de')
>>> print(li['name'], li['name_local'], li['name_translated'], li['bidi'])
German Deutsch 德语 False
```

14.4 编写模板代码

在模板中使用翻译功能需要用到两个新的模板标签：trans 和 blocktrans。为了使模板能够识别这两个标签，还需要在模板最上面添加 {% load i18n %}。需要注意的是，即使上级模板已经引用 {% load i18n %}，在当前模板中仍需要引用这条指令。

14.4.1 trans

{% trans %} 标签可以翻译字符串和变量，例如：

```
<title>{% trans "This is the title." %}</title>
<title>{% trans myvar %}</title>
```

使用 noop 参数可以跳过翻译阶段：

```
<title>{% trans "myvar" noop %}</title>
```

trans 标签内部使用 gettext() 方法，上面例子中 gettext() 方法接收字符串 "myvar"，然后去消息文件目录中查找对应的字符串。

在 {% trans %} 标签中不能使用模板变量，如果一定要使用变量，可以使用 {% blocktrans %}。

如果只是翻译一段文字并不显示它，可以将 trans 赋值给一个变量，例如：

```
{% trans "This is the title" as the_title %}

<title>{{ the_title }}</title>
<meta name="description" content="{{ the_title }}">
```

通过将翻译字符串赋值给变量，可以很方便地复用它，例如：

```
{% trans "starting point" as start %}
{% trans "end point" as end %}
{% trans "La Grande Boucle" as race %}

<h1>
  <a href="/" title="{% blocktrans %}Back to '{{ race }}' homepage{% endblocktrans %}">{{ race }}</a>
```

```
</h1>
<p>
{% for stage in tour_stages %}
    {% cycle start end %}: {{ stage }}{% if forloop.counter|divisibleby:2 %}<br>{% else %}, {% endif %}
{% endfor %}
</p>
```

通过 context 关键字，{% trans %} 标签也可以支持上下文标记，语法如下：

```
{% trans "May" context "month name" %}
```

14.4.2　blocktrans

使用 blocktrans 标签可以完成复杂的翻译工作，包括使用纯文本以及变量。语法如下：

```
{% blocktrans %}This string will have {{ value }} inside.{% endblocktrans %}
```

为了在 blocktrans 中使用模板表达式，例如访问对象属性或者使用模板过滤器，可以将模板表达式绑定给一个局部变量，例如：

```
{% blocktrans with amount=article.price %}
That will cost $ {{ amount }}.
{% endblocktrans %}

{% blocktrans with myvar=value|filter %}
This will have {{ myvar }} inside.
{% endblocktrans %}
```

在一个 blocktrans 标签中可以同时声明多个表达式，例如：

```
{% blocktrans with book_t=book|title author_t=author|title %}
This is {{ book_t }} by {{ author_t }}
{% endblocktrans %}
```

可以使用逻辑运算符使代码更容易理解，例如：

```
{% blocktrans with book|title as book_t and author|title as author_t %}
```

blocktrans 暂时不支持复杂的模板标签，如 {% for %}、{% if %} 等。如果在执行 blocktrans 代码块时失败，Django 将会调用 deactivate_all() 方法停止当前语言并使用默认语言解释剩下的代码。

blocktrans 也支持复数操作，为了启用复数，需要进行以下操作：

（1）使用名为 count 的方法声明并绑定一个计数器，该计数器决定了文本的单复数显示形式。
（2）指定文本的单复数显示格式，两种格式之间使用 {% plural %} 分隔。

例如根据调查问卷数量显示一行文本：

```
{% blocktrans count counter=latest_question_list|length %}
There is only one question.
{% plural %}
There are {{ counter }} questions.
{% endblocktrans %}
```

使用 blocktrans 的时候一定要记住，blocktrans 是使用 ngettext 对字符串进行翻译的，一切都应该遵守 ngettext() 方法的使用规则。

使用 asvar 标签可以将 blocktrans 保存为一个变量，例如：

```
{% blocktrans asvar the_title %}The title is {{ title }}.{% endblocktrans %}
<title>{{ the_title }}</title>
<meta name="description" content="{{ the_title }}">
```

以上代码将"The title is {{ title }}."保存为变量 the_title,后面可以直接使用 the_title。

在 blocktrans 中使用 context 标签,也可以实现上下文标记的效果:

```
{% blocktrans with name=user.username context "greeting" %}Hi {{ name }}{% endblocktrans %}
```

最后一个小技巧是 trimmed,在 blocktrans 中使用 trimmed 标签将会删除所有换行符,同时把每一行前后的空格删掉,最后把所有行合并为一行显示,例如:

```
{% blocktrans trimmed %}
  First sentence.
  Second paragraph.
{% endblocktrans %}
```

以上代码在网页上将会显示一行文本:

```
First sentence. Second paragraph.
```

14.4.3 注释

模板中的翻译字符串也是使用 {% comment %} 进行注释的,例如:

```
{% comment %}Translators: View verb{% endcomment %}
{% trans "View" %}

{% comment %}Translators: Short intro blurb{% endcomment %}
<p>{% blocktrans %}A multiline translatable
literal.{% endblocktrans %}</p>
```

如果是单行注释,可以使用 {# ... #}:

```
{# Translators: Label of a button that triggers search #}
<button type="submit">{% trans "Go" %}</button>

{# Translators: This is a text of the base template #}
{% blocktrans %}Ambiguous translatable block of text{% endblocktrans %}
```

14.5 翻译原理

应用程序运行时 Django 会创建一个基于内存的文本翻译目录。为了实现这一点,Django 按照以下逻辑在不同的文件路径中查找消息文件(.mo)以及确定同一文本的不同翻译的优先级:

(1)LOCALE_PATHS 中的目录具有更高的优先级,先出现在 LOCALE_PATHS 的目录比后出现在 LOCALE_PATHS 的目录优先级更高。

(2)接下来查找并使用每个应用程序目录下的 locale 文件夹,如果有,先出现的比后出现的优先级高。

(3)使用 Django 提供的翻译文件,文件路径:django/conf/locale。

在任何时候,存放翻译文件的文件夹都应该使用地区符号作为文件夹名字,如 de、pt_BR、es_AR 等。区域语言变体的未翻译字符串使用通用语言的翻译。例如,未翻译的 pt_BR 字符串使用 pt 进行翻译。

基于此,开发人员可以为应用程序开发自己的翻译文件,也可以重写 Django 提供的翻译文件。为了使项目更规范、易于管理,还可以将所有翻译文件提取到一个公共路径。

所有消息文件仓库(翻译文件仓库)都使用统一的组织规范:

- LOCALE_PATHS 中的路径都会被按照 <language>/LC_MESSAGES/django.(po|mo) 的格式进行搜索,如 zh_Hans/LC_MESSAGES/django.mo。
- $APPPATH/locale/<language>/LC_MESSAGES/django.(po|mo)。
- $PYTHONPATH/django/conf/locale/<language>/LC_MESSAGES/django.(po|mo)。

可以使用 django-admin makemessages 创建消息文件,使用 django-admin compilemessages 生成二进制文件(.mo),这些文件被 gettext() 方法使用。

使用 django-admin compilemessages --settings=path.to.settings 命令可以一次编译 LOCALE_PATHS 中的所有文件。

当准备好 .po 和 .mo 文件之后,就可以配置 Django 应用了。

配置文件中的 LANGUAGE_CODE 可以作为默认的语言设置,当 Django 无法找到更匹配的语言时将使用 LANGUAGE_CODE 所指定的语言显示网页。

如果希望每一个用户都可以设置自己的语言偏好,那么还需要激活 LocaleMiddleware 中间件。LocaleMiddleware 使得 Django 能够基于请求来确认语言,这就使得每一个用户都可以设置自己的语言偏好。

激活 LocaleMiddleware 中间件要求保证以下顺序:

(1)保证它是第一个安装的中间件之一。

(2)要放在 SessionMiddleware 之后,因为 LocaleMiddleware 中间件使用 session 数据;要放在 CommonMiddleware 之前,因为 CommonMiddleware 要使用当前语言来转义 URL。

(3)如果激活了 CacheMiddleware,要把 LocaleMiddleware 放在它后面。

配置好的 MIDDLEWARE 可能是下面这个样子:

```
MIDDLEWARE = [
    'django.contrib.sessions.middleware.SessionMiddleware',
    'django.middleware.locale.LocaleMiddleware',
    'django.middleware.common.CommonMiddleware',
]
```

LocaleMiddleware 中间件将根据以下顺序判断用户的语言偏好:

(1)它会在请求的 URL 中查找语言前缀。这只有在根 URL 中使用 i18n_patterns 函数时才会进行。

(2)查找 Cookie。对应的 Cookie 名字在 LANGUAGE_COOKIE_NAME 中进行设置,默认值为 django_language。

(3)查找 HTTP 头的 Accept-Language。这个头是由浏览器发送的,将会通知服务器用户的语言偏好,以及优先级顺序。Django 按顺序尝试每种语言,直到找到一个可用的语言。

(4)如果以上尝试都没有找到一个合适的语言,最后将使用全局的 LANGUAGE_CODE 配置项。

> **注意**
>
> 在某些地方，语言首选项是满足标准语言格式的一个字符串。例如，巴西葡萄牙语是pt-br。
>
> 如果基本语言可用，但指定的子语言不可用，Django使用基本语言。例如，如果用户指定de-at（奥地利德语），但Django只有德语（de）可用，那么Django使用德语。
>
> 只能选择在LANGUAGES配置项中列举的语言。例如：
>
> ```
> LANGUAGES = (
>     ('en-us', ('English')),
>     ('zh-CN', (' 中文简体 ')),
> )
> ```

当配置好LocaleMiddleware中间件之后，用户语言偏好就会被设置到request.LANGUAGE_CODE中。此时不同地区的用户访问网站时翻译字符串将会按照用户的语言偏好进行显示，例如中文时区的用户访问调查问卷系统的效果如下图所示。

第 15 章

安全

由于互联网是一个开放的环境,在网络上存在各种各样的危险,因此在将 Django 应用程序部署到 Web 服务器之前需要检查应用程序的配置信息,包括安全、性能以及其他选项,保证 Django 应用能够以最优的状态在互联网中运行。

15.1 网络攻击与保护

Django 提供了对多种网络攻击的防护能力,本节介绍部分网络攻击类型以及 Django 对应的保护方式。

15.1.1 跨站脚本攻击

跨站脚本攻击(Cross Site Scripting,XSS),指黑客在用户访问的网页中注入恶意代码或者引诱用户单击特定的网络连接而执行攻击脚本,从而达到攻击目的。这些攻击脚本可能来自黑客上传到数据库的一段脚本,当网站显示这些数据时就可能诱导用户单击。

Django 模板会将以下 5 个变量标签进行安全转义:

- < 被替换为 <。
- \> 被替换为 >。
- '(单引号)被替换为 '。
- "(双引号)被替换为 "。
- & 被替换为 &。

因此,当黑客在网页中注入恶意代码时,代码将不会被执行。

例如,博客文章 5 在数据库中的存储形式如下图所示。

id	name	tagline
1	Cheddar Talk	All the latest Beatles news.
2	Aarons Blog	
3	Awsome Django	Django web framework is awsome
4	MTV Framework	Django uses MTV framework
5	Attack	\<script src="攻击脚本.js"\>\</script\>

当用户浏览到该文章时,Django 模板会对其进行转义,如下图所示。

```
23  <div class="center">
24      <h2 class="blog_head">5 - Attack</h2>
25      <p class="blog_body">
26          &lt;script src=&quot;攻击脚本.js&quot;&gt;&lt;/script&gt;
27      </p>
28  </div>
```

在浏览器上仍然能正常显示，效果如下图所示。

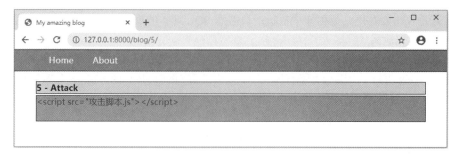

但是，并不是所有特殊字符都需要转义，此时可以使用 safe 过滤器将字符标记为安全字符。例如，当变量 data 是"<b>"时，以下两种写法将会产生截然不同的输出结果：

```
This will be escaped: {{ data }}
This will not be escaped: {{ data|safe }}
```

输出结果如下：

```
This will be escaped: &lt;b&gt;
This will not be escaped: <b>
```

safe 过滤器适用于单一变量的转义，如果代码段包含多个变量需要进行安全设置时，可以使用 autoescape 标签、autoescape 标签接收 on 或者 off 参数，on 表示启用模板的强制转换功能，off 表示禁用强制转换。以下是 autoescape 的简单应用：

```
{% autoescape off %}
    This will not be auto-escaped: {{ data }}.

    Nor this: {{ other_data }}
    {% autoescape on %}
        Auto-escaping applies again: {{ name }}
    {% endautoescape %}
{% endautoescape %}
```

需要注意的是，由于 Django 模板的变量值通常是从数据库中读取的，因此在保存数据时一定要注意数据的安全，尤其是在保存 HTML 代码时。

15.1.2 跨站请求伪造攻击

跨站请求伪造攻击（Cross Site Request Forgery，CSRF），攻击者会在用户不知情的情况下使用用户的安全证书进行非法攻击。例如，当用户登录网银后，在没有退出系统的情况下访问了攻击者的恶意网站，此时恶意网站可以利用用户已有的 session 进行非法操作。

在前面的学习中已经接触了 Django 的 CSRF 保护。CSRF 保护可以保护用户免受绝大多数的 CSRF 攻击，但是仍然存在一些限制，如开发人员人为地在全站或者特定视图中禁用 CSRF 模块，又或者网站存在一个全新的子站，主站与子站处于不同域。

Django 的 CSRF 保护会检查每一个 POST 请求，这使得攻击者不能通过简单的提交表单的方式进行非法操作。

如果使用 HTTPS 协议部署网站，CsrfViewMiddleware 将会检查每一个 HTTP 报文的 referer 头是否来自同一个域（包括子域和端口号）。

15.1.3 SQL 注入

SQL 注入是攻击者在网站数据库中执行一段恶意 SQL 脚本的攻击方式，这类攻击通常会导致数据库被删除或者数据泄漏。

Django 的 queryset 能够有效地阻止 SQL 注入攻击，由于 SQL 脚本中的参数可能来自用户提交的数据，因此 queryset 将每一个参数都进行转义，这样保证了任何被执行的 SQL 脚本都是安全可靠的。

由于 Django 给予开发人员很大的自由空间，因此开发人员仍然可以编写自定义 SQL 脚本，对于这种情况，开发人员一定要注意代码的安全。

15.1.4 点击劫持

点击劫持是在恶意网站中嵌入一个 iframe 的方式诱使用户点击，从而达到非法入侵的网络攻击方式，制造这种攻击的成本高，非常少见。

Django 利用中间件 XFrameOptionsMiddleware 可以有效地阻止自己的网站被其他网站以 frame 的方式引用。

如果网站不需要在 frame 中引用，强烈建议启用 XFrameOptionsMiddleware。

在 settings 中引用 XFrameOptionsMiddleware：

```
MIDDLEWARE = [
    ...
    'django.middleware.clickjacking.XFrameOptionsMiddleware',
    ...
]
```

15.2 检查配置信息

可以使用 manage.py check --deploy 命令检查当前配置信息是否存在安全隐患。下图是本书示例代码的检查结果，从输出结果可见，当前系统存在 8 个安全隐患。

接下来介绍安全检查项。

1. SECRET_KEY

SECRET_KEY 必须是一个足够长的随机字符串。SECRET_KEY 只能在一个网站中使用，绝对不能泄漏。

除了直接在 settings 中设置 SECRET_KEY 外，还可以在环境变量中加载它：

```
import os
SECRET_KEY = os.environ['SECRET_KEY']
```

也可以在文件中加载它：

```
with open('/etc/secret_Key.txt') as f:
    SECRET_KEY = f.read().strip()
```

2. DEBUG

DEBUG 配置只可用于开发环境，在生产环境中绝对不能启用，因为 DEBUG 配置将会把所有代码跟踪信息显示到网页中，这些信息包含很多敏感内容，如数据库信息等。

3. ALLOWED_HOSTS

设置 Django 应用的主机名，只有访问 ALLOWED_HOSTS 中指定主机的请求才可以被 Django 处理。当设置 DEBUG = False 时，必须设置 ALLOWED_HOSTS，否则 Django 不允许任何请求访问网站。这个配置可以有效地阻止 CSRF 攻击。

4. CACHES

如果 Django 应用程序使用了缓存，那么开发环境与生产环境通常会引用不同的缓存地址，因此在部署 Django 应用程序时要检查缓存地址是否正确。

5. DATABASES

如果 Django 应用程序使用了数据库，那么开发环境与生产环境通常会使用不同的数据库，因此在部署 Django 应用程序时要检查数据库配置信息是否正确。

6. EMAIL_BACKEND

如果系统使用了邮件功能，那么需要保证在部署到生产环境时使用正确的配置。默认情况下 Django 使用 webmaster@localhost 和 root@localhost 发送邮件，如果希望使用其他邮件地址需要修改 DEFAULT_FROM_EMAIL 和 SERVER_EMAIL。

7. STATIC_ROOT 和 STATIC_URL

在开发环境中系统默认使用开发机器管理静态文件，但是在生产环境会使用其他地址，因此需要重新定义 collectstatic 能够访问的 STATIC_ROOT 路径。

8. MEDIA_ROOT 和 MEDIA_URL

由于媒体文件是用户上传的文件，因此这些文件是不安全文件，一定要保证 Django 系统不能够与这些文件进行交互，例如用户上传了一个 .py 文件，这个文件很可能会泄漏系统数据或破坏系统安全。

9. CSRF_COOKIE_SECURE 和 SESSION_COOKIE_SECURE

在生产环境中需要将这两个配置设置为 true，以防止意外通过 HTTP 传递 CSRF Cookie 和 session Cookie。

10. CONN_MAX_AGE

表示数据库连接的生命周期，如果将 CONN_MAX_AGE 设置为 0，系统会在每个请求结束后自动关闭连接；如果设置为 None，数据库连接将成为一个持久连接，虽然这样会提高数据库访问速度，但是会持续占有数据库资源，需要根据网站实际情况进行设置。

第 16 章 部署

16.1 WSGI 和 Application 对象

WSGI 是 Web Server Gateway Interface 的缩写，用于描述 Web 服务器与 Web 应用程序之间的通信方式，是 Django 应用程序的最主要部署平台。通过使用 startproject 命令创建 Django 工程的时候会自动创建一个简单的 WSGI 配置文件（wsgi.py）。

用 WSGI 部署的关键是 application callable，Web 服务器用它与用户的代码进行交互。在 wsgi.py 脚本中使用 get_wsgi_application() 方法获取一个 application 对象实例。

16.2 Ubuntu 部署 Django

16.2.1 查看系统版本

输入如下命令，查看系统版本。

```
$ lsb_release -a
```

结果如下图所示。

```
aaron@ubuntu:~$ lsb_release -a
No LSB modules are available.
Distributor ID: Ubuntu
Description:    Ubuntu 18.04.1 LTS
Release:        18.04
Codename:       bionic
```

16.2.2 更换国内源

由于系统自带的源是国外服务器，因此安装软件时可能会很慢，可以考虑切换成国内源。
首先备份原来的源：

```
$ sudo cp /etc/apt/sources.list /etc/apt/sources.list.bak
```

编辑源：

```
$ sudo vi /etc/apt/sources.list
```

将 source.list 文件内容替换成下面清华大学的源地址：

```
    deb https://mirrors.tuna.tsinghua.edu.cn/ubuntu/ bionic main restricted universe
multiverse
    deb https://mirrors.tuna.tsinghua.edu.cn/ubuntu/ bionic-updates main restricted
universe multiverse
    deb https://mirrors.tuna.tsinghua.edu.cn/ubuntu/ bionic-backports main restricted
universe multiverse
    deb https://mirrors.tuna.tsinghua.edu.cn/ubuntu/ bionic-security main restricted
universe multiverse
    deb https://mirrors.tuna.tsinghua.edu.cn/ubuntu/ bionic-proposed main restricted
universe multiverse
    deb-src https://mirrors.tuna.tsinghua.edu.cn/ubuntu/ bionic main restricted
universe multiverse
    deb-src https://mirrors.tuna.tsinghua.edu.cn/ubuntu/ bionic-updates main restricted
universe multiverse
    deb-src https://mirrors.tuna.tsinghua.edu.cn/ubuntu/ bionic-backports main
restricted universe multiverse
    deb-src https://mirrors.tuna.tsinghua.edu.cn/ubuntu/ bionic-security main
restricted universe multiverse
    deb-src https://mirrors.tuna.tsinghua.edu.cn/ubuntu/ bionic-proposed main
restricted universe multiverse
```

执行以下命令更新源文件索引：

```
$ sudo apt-get update
```

更新软件：

```
$ sudo apt-get upgrade
```

国内还有其他优秀的源可用，读者可以自行选择。

16.2.3 查看 Python 版本

输入如下命令，查看 Python 版本。

```
$ python3 -V
```

结果如下图所示。

> **注意**
>
> 本书的演示系统默认已经安装了 Python 3，可以使用 python3 命令查看版本。很多系统默认安装的是 Python 2，执行版本检查时使用 python -V 命令。

由于本书使用 Python 3.8 作为演示环境，因此需要升级 Python，执行以下命令：

```
$ sudo apt install python3.8
```

安装结束查看 Python 3.8 的安装路径：

```
$ whereis python3.8
```

结果如下图所示。

```
aaron@ubuntu:~$ whereis python3.8
python3: /usr/bin/python3.6m /usr/bin/python3 /usr/bin/python3.8 /usr/bin/pytho
n3.6 /usr/lib/python3 /usr/lib/python3.7 /usr/lib/python3.8 /usr/lib/python3.6
/etc/python3 /etc/python3.8 /etc/python3.6 /usr/local/lib/python3.8 /usr/local/
lib/python3.6 /usr/include/python3.6m /usr/share/python3 /usr/share/man/man1/py
thon3.1.gz
```

此时 Python 3.8 已经被安装在 /usr/bin/ 目录中，如下图所示。

```
aaron@ubuntu:/etc/python3.8$ /usr/bin/python3.8
Python 3.8.0 (default, Oct 28 2019, 16:14:01)
[GCC 8.3.0] on linux
Type "help", "copyright", "credits" or "license" for more information.
```

为 Python 3.8 建立软链接：

```
$ sudo ln -s /usr/bin/python3.8 /usr/bin/python
```

此时可以使用 python-V 命令查看 python 版本，如下图所示。

```
aaron@ubuntu:~$ python -V
Python 3.8.0
```

16.2.4　安装 pip3

执行以下命令安装 pip3：

```
$ sudo apt-get install python3-pip
```

安装完 pip3 之后同样设置软链接：

```
$ whereis pip3
pip3: /usr/bin/pip3 /usr/share/man/man1/pip3.1.gz
$ sudo cp /usr/bin/pip3 /usr/bin/pip3.bak
$ sudo ln -s /usr/bin/pip3 /usr/bin/pip
```

查看 pip 文件内容，如下图所示。

```
aaron@ubuntu:~$ cat /usr/bin/pip
#!/usr/bin/python3
# GENERATED BY DEBIAN

import sys

# Run the main entry point, similarly to how setuptools does it, but because
# we didn't install the actual entry point from setup.py, don't use the
# pkg_resources API.
from pip import main
if __name__ == '__main__':
    sys.exit(main())
```

可以看到 pip 命令使用的解释器是 Python 3，此时需要重新指定 Python：

```
$ sudo vi /usr/bin/pip
```

修改后的 pip 文件内容如下图所示。

```
aaron@ubuntu:~$ cat /usr/bin/pip
#!/usr/bin/python
# GENERATED BY DEBIAN

import sys

# Run the main entry point, similarly to how setuptools does it, but because
# we didn't install the actual entry point from setup.py, don't use the
# pkg_resources API.
from pip import main
if __name__ == '__main__':
    sys.exit(main())
aaron@ubuntu:~$ pip -V
pip 9.0.1 from /usr/lib/python3/dist-packages (python 3.8)
```

pip 默认的源也是国外地址，为了提升访问速度，可以更换为国内源，更换 pip 源有以下两种方式。

方式 1：临时切换源地址，如临时使用豆瓣源安装 Django：

```
$ pip install -i http://pypi.douban.com/simple --trusted-host pypi.douban.com django
```

方式 2：永久替换国内源。

首先在用户的家目录中创建一个 .pip 文件夹，在 .pip 文件夹中创建 pip.conf 文件，使用 vi 编辑器输入以下内容：

```
[global]
index-url = https://pypi.douban.com/simple
[install]
use-mirrors = true
mirrors = https://pypi.douban.com/simple
trusted-host = pypi.douban.com
```

输入命令，结果如下图所示。

16.2.5　安装 nginx

执行以下命令安装 nginx：

```
$ sudo apt install nginx
```

安装结束可以在浏览器中查看 nginx 欢迎页，如下图所示。

16.2.6　安装 Django

输入如下命令，安装 Django。

```
$ pip install Django
```

16.2.7　安装 uwsgi

安装 uwsgi 之前先安装依赖包：

```
$ sudo apt-get install build-essential python3.8-dev
```

> **注意**
>
> 依赖包要与 Python 版本严格匹配，即 Python 3.8 必须使用 python3.8-dev。

执行以下命令安装 uwsgi：

```
$ pip install uwsgi
```

16.2.8 命令行运行网站

首先需要将 Django 代码复制到 /var/www/html/mysite 目录，复制后的目录如下图所示。

使用 runserver 命令运行 Django 网站：

```
$ python manage.py runserver 你的服务器IP地址
```

访问网站时可能会出现以下错误：

```
Invalid HTTP_HOST header: '192.168.1.103:8080'. You may need to add '192.168.1.103' to ALLOWED_HOSTS.
```

这个错误是由于配置文件没有设置导致的。打开 settings.py，修改 DEBUG 以及 ALLOWED_HOSTS，如下图所示。

```
# SECURITY WARNING: don't run with debug turned on in production!
DEBUG = False

ALLOWED_HOSTS = ["127.0.0.1","192.168.1.103"]
```

重新运行网站，此时能够正常访问了，说明网站代码以及 Django 环境都已经设置好。

16.2.9 配置 uwsgi

在 manage.py 文件的相同目录创建一个 uwsgi.ini 文件，uwsgi.ini 文件可以包括以下内容：

```
[uwsgi]
# 直接访问 uwsgi 的端口号，绕过 nginx
http = :8010
# 转发给 nginx 的端口号
socket = 127.0.0.1:8080
# 是否使用主线程
master = true
```

```
# 项目的绝对路径
chdir = /var/www/<PROJECT_NAME>/
# Django 项目 wsgi.py 文件的相对路径
wsgi-file = <PROJECT_NAME>/wsgi.py
# 进程数
processes = 4
# 每个进程的线程数
threads = 2
# 监听端口
stats = 127.0.0.1:9191
# 每次退出时是否清理环境配置
vacuum = true
# 目录中一旦有文件被改动就自动重启
touch-reload = /var/www/html/mysite
# 存放日志
daemonize = /var/www/my_site/uWSGI.log
```

本书以最简单的 uwsgi 配置为例部署网站，如下图所示。

配置完成后可以执行以下命令检查 uwsgi 是否配置成功：

```
$ uwsgi --ini uwsgi.ini
```

```
aaron@ubuntu:/var/www/html/mysite$ cat uwsgi.ini
[uwsgi]
socket = 127.0.0.1:8080
chdir = /var/www/html/mysite
wsgi-file = mysite/wsgi.py
processes = 4
threads = 2
#stats = 127.0.0.1:9191
aaron@ubuntu:/var/www/html/mysite$
```

16.2.10　配置 nginx

nginx 配置文件路径：/etc/nginx/nginx.conf。

在 http 节点增加 server 节点，如下图所示。

```
http {
    server {
        listen 80;
        server_name 192.168.1.103;
        charset utf-8;
        location / {
            # 必须和uwsgi.ini中socket一样
            uwsgi_pass 127.0.0.1:8080;
            include uwsgi_params;
        }
    }
}
```

配置完 nginx 需要重启服务器。

16.2.11　启动网站

启动 uwsgi：

```
$ uwsgi --ini uwsgi.ini
```

启动 nginx：

```
$ sudo nginx -c /etc/nginx/nginx.conf
```

此时使用服务器 IP 地址（nginx 配置文件中的 server_name）就可以在局域网内任意一台电脑上访问网站了，如下图所示。

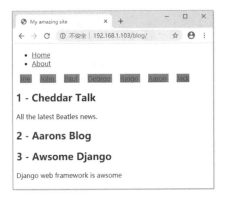

16.2.12　设置静态文件

网站已经可以运行了，下面调整静态文件使网站能够更好地展示。

首先创建一个静态文件路径，将该路径添加到 settings.py：

```
STATIC_ROOT = '/var/www/html/mysite/static'
```

执行 collectstatic 命令将全部静态文件复制到该路径：

```
# python manage.py collectstatic
```

修改 nginx.conf 文件，添加对静态文件的路由，如下图所示。

```
server {
    listen 80;
    server_name 192.168.1.103;
    charset utf-8;
    location /static {
        alias /var/www/html/mysite/static;
    }
    location / {
        # 必须和uwsgi.ini中socket一样
        uwsgi_pass 127.0.0.1:8080;
        include uwsgi_params;
    }
}
```

最后重启 nginx，重启服务器，访问网站，如下图所示。

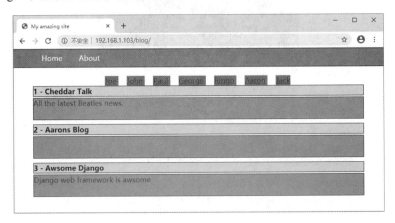

16.2.13 自启动服务

虽然网站运行起来了，但是，当用户断开服务器连接或者 uwsgi 命令行窗口关闭时，网站服务也就停止了，为了使网站能够稳定运行，还需要将网站以系统服务的方式运行。

首先创建一个服务文件：

```
$ sudo vi /etc/systemd/system/mysite.service
```

> 文件名将会被用作服务名，这里的服务名就是 mysite。

在服务文件中输入下图所示内容。

```
aaron@ubuntu:~$ cat /etc/systemd/system/mysite.service
[Unit]
Description=uWSGI instance to serve summary
After=network.target
[Service]
User=aaron
Group=www-data
WorkingDirectory=/var/www/html/mysite
ExecStart=/home/aaron/.local/bin/uwsgi /var/www/html/mysite/uwsgi.ini
[Install]
WantedBy=multi-user.target
aaron@ubuntu:~$
```

配置项说明如下。

User：启动网站服务的账号。
Group：用户组。
WorkingDirectory：网站文件路径。
ExecStart：启动 uwsgi 服务的命令，uwsgi 的位置可以使用 whereis 命令获得。

关于系统服务的常用命令：

停止服务：sudo systemctl stop mysite。
启动服务：sudo systemctl start mysite。
重启服务：sudo systemctl restart mysite。
重新加载 service 文件：systemctl daemon-reload。

此时网站就已经能随着服务器启动自动运行了。

16.3 CentOS 部署 Django

16.3.1 查看系统版本

输入如下命令，查看 CentOS 系统版本。

```
[root@localhost ~]# cat /etc/centos-release
CentOS Linux release 7.5.1804 (Core)
[root@localhost ~]#
```

16.3.2 更换国内源

以清华大学镜像源为例演示如何更新 CentOS 源。

首先备份本地 CentOS-Base.repo：

```
# sudo cp /etc/yum.repos.d/CentOS-Base.repo /etc/yum.repos.d/CentOS-Base.repo.bak
```

然后编辑 /etc/yum.repos.d/CentOS-Base.repo 文件，在"mirrorlist="开头行前面加 # 注释掉，并将以"baseurl="开头的行取消注释（如果被注释了），把该行内的域名（例如 mirror.centos.org）替换为 https://mirrors.tuna.tsinghua.edu.cn。

清除 yum 缓存：

```
# yum clean all
```

更新软件包缓存：

```
# sudo yum makecache
```

更新 yum：

```
# yum update
```

16.3.3 更新 Python

系统默认的 Python 版本为 2.7，因此需要更新到 Python 3.8。

```
[root@bogon ~]# python -V
Python 2.7.5
[root@bogon ~]#
```

安装 wget 工具：

```
# yum install -y wget
```

安装 gcc：

```
# yum install gcc
```

由于 Django 3.x 要求的 SQLite 最低版本是 3.8.3，而系统自带的 SQLite 版本过低（可以通过 sqlite3-version 命令查看系统 SQLite 版本），因此，在安装 Python 时需提前安装最新版本的 SQLite，否则运行 Django 应用程序的时候就会出现以下错误：

```
django.core.exceptions.ImproperlyConfigured: SQLite 3.8.3 or later is required (found 3.7.17).
```

升级 SQLite 的代码如下：

```
# wget https://www.sqlite.org/2020/sqlite-autoconf-3310100.tar.gz
# tar zxvf sqlite-autoconf-3310100.tar.gz
# cd sqlite-autoconf-3310100
# ./configure --prefix=/usr/local/sqlite
# make && make install
```

> ./configure --prefix=/usr/local/sqlite 命令指定 SQLite 的安装目录，安装 Python 3.8 的时候需要使用这个路径。

接下来安装 Python 3.8。

安装依赖：

```
# yum install zlib-devel bzip2-devel openssl-devel ncurses-devel sqlite-devel readline-devel tk-devel libffi-devel make
```

下载最新稳定版 Python：

```
# wget https://www.python.org/ftp/python/3.8.3/Python-3.8.3.tgz
```

解压 Python 包：

```
# tar -zxvf Python-3.8.3.tgz
# cd Python-3.8.3
```

编辑安装 Python 3.8：

```
# LD_RUN_PATH=/usr/local/sqlite/lib ./configure LDFLAGS="-L/usr/local/sqlite/lib" CPPFLAGS="-I/usr/local/sqlite/include"  --prefix=/usr/local/python3.8
# LD_RUN_PATH=/usr/local/sqlite/lib make
# LD_RUN_PATH=/usr/local/sqlite/lib sudo make install
```

> 必须严格按照以上步骤安装 Python，否则可能仍然使用低版本 SQLite。

建立软链接：

```
# cd /usr/bin/
# sudo mv python python.bak
# sudo ln -s /usr/local/python3.8/bin/python3.8 /usr/bin/python
# sudo mv pip pip.bak
# sudo ln -s /usr/local/python3.8/bin/pip3 /usr/bin/pip
```

验证 Python 环境，如下图所示。

```
[root@bogon bin]# python -V
Python 3.8.3
[root@bogon bin]# pip -V
pip 19.2.3 from /usr/local/lib/python3.8/site-packages/pip (python 3.8)
[root@bogon bin]#
```

安装完毕，参考 16.2.4 节更新 pip 国内源。

由于 yum 命令使用的是 Python 2，因此需要重新指定 Python 版本，否则会出现类似下图所示的错误。

```
[root@bogon .pip]# yum install -y epel-release
  File "/usr/bin/yum", line 30
    except KeyboardInterrupt, e:
                            ^
SyntaxError: invalid syntax
```

使用 vi 编辑器打开 yum 文件：

```
# vi /usr/bin/yum
```

修改第一行代码：

```
#!/usr/bin/python2.7
```

使用 vi 编辑器打开 urlgrabber-ext-down 文件：

```
# cd /usr/libexec/
# vi urlgrabber-ext-down
```

将 #/usr/bin/python 改为 #/usr/bin/python2.7。

16.3.4 安装 Django

输入如下命令安装 Django。

```
# pip install django
```

16.3.5 安装 uwsgi

输入如下命令安装 uwsgi。

```
# yum install python-devel
# python -m pip install uwsgi
# find / -name uwsgi
/usr/local/python3.8/bin/uwsgi
# ln -s /usr/local/python3.8/bin/uwsgi /usr/bin/uwsgi
```

16.3.6 命令行运行网站

将项目代码复制到 /var/www/html/mysite，执行 runserver 命令，运行成功，如下图所示。

```
[root@bogon mysite]# python manage.py runserver
Performing system checks...

System check identified no issues (0 silenced).
May 19, 2020 - 04:23:19
Django version 3.0.6, using settings 'mysite.settings'
Starting development server at http://127.0.0.1:8000/
Quit the server with CONTROL-C.
```

16.3.7 配置 uwsgi

在 manage.py 文件的相同目录创建 uwsgi.ini 文件，文件内容如下图所示。

```
[root@bogon mysite]# cat uwsgi.ini
[uwsgi]
socket = 127.0.0.1:8080
chdir = /var/www/html/mysite
wsgi-file = mysite/wsgi.py
processes = 4
threads = 2
```

16.3.8 安装 nginx

执行以下命令安装 nginx：

```
# yum install -y epel-release
# yum -y install nginx
```

安装完重启服务器。

配置 nginx：

```
# cd /etc/nginx/
```

按照下图所示方式修改 nginx.conf 的 http 节点下的 server 节点。

```
server {
    listen       80 default_server;
    listen       [::]:80 default_server;
    server_name  192.168.1.104;
    root         /usr/share/nginx/html;

    # Load configuration files for the default server block.
    include /etc/nginx/default.d/*.conf;

    location / {
        uwsgi_pass 127.0.0.1:8080;
        include uwsgi_params;
    }

    error_page 404 /404.html;
        location = /40x.html {
    }

    error_page 500 502 503 504 /50x.html;
        location = /50x.html {
    }
}
```

如果网络上其他机器不能访问 nginx，可能是防火墙阻拦了，设置防火墙过滤规则即可：

```
# firewall-cmd --zone=public --add-port=80/tcp --permanent
# firewall-cmd --reload
```

注意

--add-port=80/tcp 对应 nginx 对外开放的端口号。

启动 uwsgi：

```
# uwsgi --ini uwsgi.ini
```

启动 nginx：

```
# nginx -c /etc/nginx/nginx.conf
```

此时在局域网内已经可以访问 Django 网站了，但是样式文件仍然没有加载成功，需要配置静态文件。

与 16.2 节一样配置静态文件。最终 nginx 的配置信息如下图所示。

```nginx
server {
    listen       80 default_server;
    listen       [::]:80 default_server;
    server_name  192.168.1.104;
    root         /usr/share/nginx/html;

    # Load configuration files for the default server block.
    include /etc/nginx/default.d/*.conf;

    location /static {
        alias /var/www/html/mysite/static;
    }

    location / {
        uwsgi_pass 127.0.0.1:8080;
        include uwsgi_params;
    }

    error_page 404 /404.html;
        location = /40x.html {
    }

    error_page 500 502 503 504 /50x.html;
        location = /50x.html {
    }
}
```

16.3.9　自启动服务

与 Ubuntu 系统一样需要设置网站开机自启。

首先在 /usr/lib/systemd/system/ 目录创建服务文件 mysite.service，文件内容如下图所示。

```
[root@bogon system]# cat mysite.service
[Unit]
Description=uWSGI instance to serve summary
After=network.target

[Service]
KillSignal=SIGQUIT
Restart=always
Type=notify
NotifyAccess=all
ExecStart=/usr/bin/uwsgi --ini /var/www/html/mysite/uwsgi.ini
PrivateTmp=true

[Install]
WantedBy=multi-user.target
```

最后将服务加入 systemd 中：

```
# systemctl start mysite.service
```

此时重启服务器或者执行以下命令就可以启动网站，不再需要单独启动 uwsgi：

```
# systemctl start mysite.service
```

附录 A
语言码

语言码	语言名	语言码	语言名
af	南非荷兰语	fa	波斯语
ar	阿拉伯语	fi	芬兰语
ast	阿斯图里亚斯语	fr	法语
az	阿塞拜疆语	fy	弗里西亚语
bg	保加利亚语	ga	爱尔兰语
be	白俄罗斯语	gd	苏格兰盖尔语
bn	孟加拉语	gl	加利西亚语
br	布列塔尼语	he	希伯来语
bs	波斯尼亚语	hi	印地语
ca	加泰罗尼亚语	hr	克罗地亚语
cs	捷克语	hsb	上索布语
cy	威尔士语	hu	匈牙利语
da	丹麦语	hy	亚美尼亚语
de	德语	ia	国际语
dsb	下索布语	id	印度尼西亚语
el	希腊语	io	伊多语
en	英语	is	冰岛语
en-au	澳大利亚英语	it	意大利语
en-gb	英式英语	ja	日语
eo	世界语	ka	格鲁吉亚语
es	西班牙语	kab	卡拜尔语
es-ar	阿根廷西班牙语	kk	哈萨克语
es-co	哥伦比亚西班牙语	km	高棉语
es-mx	墨西哥西班牙语	kn	坎那达语

续表

语言码	语言名	语言码	语言名
es-ni	尼加拉瓜西班牙语	ko	朝鲜语
es-ve	委内瑞拉西班牙语	lb	卢森堡语
et	爱沙尼亚语	lt	立陶宛语
eu	巴斯克语	lv	拉脱维亚语
mk	马其顿语	sl	斯洛文尼亚语
ml	马拉雅拉姆语	sq	阿尔巴尼亚语
mn	蒙古语	sr	塞尔维亚语
mr	马拉地语	sr-latn	塞尔维亚拉丁语
my	缅甸语	sv	瑞典语
nb	挪威书面挪威语	sw	斯瓦希里语
ne	尼泊尔语	ta	泰米尔语
nl	荷兰语	te	泰卢固语
nn	挪威尼诺斯克语	th	泰国语
os	奥塞梯语	tr	土耳其语
pa	旁遮普语	tt	鞑靼语
pl	波兰语	udm	乌德穆尔特语
pt	葡萄牙语	uk	乌克兰语
pt-br	巴西葡萄牙语	ur	乌尔都语
ro	罗马尼亚语	vi	越南语
ru	俄语	zh-hans	简体中文
sk	斯洛伐克语	zh-hant	繁体中文

附录 B
日期格式化字符串

格式化字符		描述	输出
Day	d	Day of the month, 2 digits with leading zeros.	'01' to '31'
	j	Day of the month without leading zeros.	'1' to '31'
	D	Day of the week, textual, 3 letters.	'Fri'
	l	Day of the week, textual, long.	'Friday'
	S	English ordinal suffix for day of the month, 2 characters.	'st', 'nd', 'rd' or 'th'
	w	Day of the week, digits without leading zeros.	'0' (Sunday) to '6' (Saturday)
	z	Day of the year.	1 to 366
Week	W	ISO-8601 week number of year, with weeks starting on Monday.	1, 53
Month	m	Month, 2 digits with leading zeros.	'01' to '12'
	n	Month without leading zeros.	'1' to '12'
	M	Month, textual, 3 letters.	'Jan'
	b	Month, textual, 3 letters, lowercase.	'jan'
	E	Month, locale specific alternative representation usually used for long date representation.	'listopada' (for Polish locale, as opposed to 'Listopad')
	F	Month, textual, long.	'January'
	N	Month abbreviation in Associated Press style. Proprietary extension.	'Jan.', 'Feb.', 'March', 'May'
	t	Number of days in the given month.	28 to 31
Year	y	Year, 2 digits.	'99'
	Y	Year, 4 digits.	'1999'
	L	Boolean for whether it's a leap year.	True or False
	o	ISO-8601 week-numbering year, corresponding to the ISO-8601 week number (W) which uses leap weeks. See Y for the more common year format.	'1999'

续表

格式化字符		描述	输出
Time	g	Hour, 12-hour format without leading zeros.	'1' to '12'
	G	Hour, 24-hour format without leading zeros.	'0' to '23'
	h	Hour, 12-hour format.	'01' to '12'
	H	Hour, 24-hour format.	'00' to '23'
	i	Minutes.	'00' to '59'
	s	Seconds, 2 digits with leading zeros.	'00' to '59'
	u	Microseconds.	000000 to 999999
	a	'a.m.' or 'p.m.' (Note that this is slightly different than PHP's output, because this includes periods to match Associated Press style.)	'a.m.'
	A	'AM' or 'PM'.	'AM'
	f	Time, in 12-hour hours and minutes, with minutes left off if they're zero. Proprietary extension.	'1', '1:30'
	P	Time, in 12-hour hours, minutes and 'a.m.'/'p.m.', with minutes left off if they're zero and the special-case strings 'midnight' and 'noon' if appropriate. Proprietary extension.	'1 a.m.', '1:30 p.m.', 'midnight', 'noon', '12:30 p.m.'
Timezone	e	Timezone name. Could be in any format, or might return an empty string, depending on the datetime.	'', 'GMT', '-500', 'US/Eastern', etc.
	I	Daylight Savings Time, whether it's in effect or not.	'1' or '0'
	O	Difference to Greenwich time in hours.	'+0200'
	T	Time zone of this machine.	'EST', 'MDT'
	Z	Time zone offset in seconds. The offset for timezones west of UTC is always negative, and for those east of UTC is always positive.	-43200 to 43200
Date/Time	c	ISO 8601 format. (Note: unlike others formatters, such as "Z", "O" or "r", the "c" formatter will not add timezone offset if value is a naive datetime (see datetime.tzinfo).	2008-01-02T10:30:00.000123+02:00, or 2008-01-02T10:30:00.000123 if the datetime is naive
	r	RFC 5322 formatted date.	'Thu, 21 Dec 2000 16:01:07 +0200'
	U	Seconds since the Unix Epoch (January 1 1970 00:00:00 UTC).	